AF680239

Over the years...

Over the past 15 years and countless issues of Science to Sage magazine, I've had the privilege of meeting some of the most fascinating explorers, scientists, and visionaries. Each has offered me glimpses into the unseen world—a dimension of water that transcends our everyday understanding, beckoning us to explore its mysteries.

This book is an invitation into uncommon knowledge and hidden realms, where science and wisdom converge to reveal the elegance of creation. Here, we catch sight of the divine plan intricately woven into every molecule. Water is far more than an element; it is the conductor of life's symphony, a fluid storyteller reflecting nature's grand design and our own unique role within it. Through these pages, I invite you to journey into discoveries that illuminate water's profound connection to our existence and to creation itself. Water is not just a vessel; it resonates, holds memory, and responds to every vibration, shaping itself to the energy it encounters. In this way, we each carry an ocean within us—a personal water chamber, alive with resonance and reflection. By embracing this understanding, we step closer to realizing the harmony between science and spirit, uncovering the essence of free will, choice, and creation.

This journey also calls us to become better stewards of our planet and to honor our interconnectedness with all life.

May these explorations awaken wonder and inspire awe as we dive into the mysteries water so generously reveals—flowing as a bridge between the known and the unknown, the tangible and the divine.

Karen Elkins

Editor of *Science to Sage Magazine* and Author of the Award-Winning Book I*nsideOut: Revealing the Mysteries of Creation and the Wisdom to Live Your Life Consciously Connected*
www.sciencetosage.com
echosofwater.com
Edition 1. January 20245

> *"Water is the driving force of all nature."*
> — Leonardo da Vinci

CONTENTS

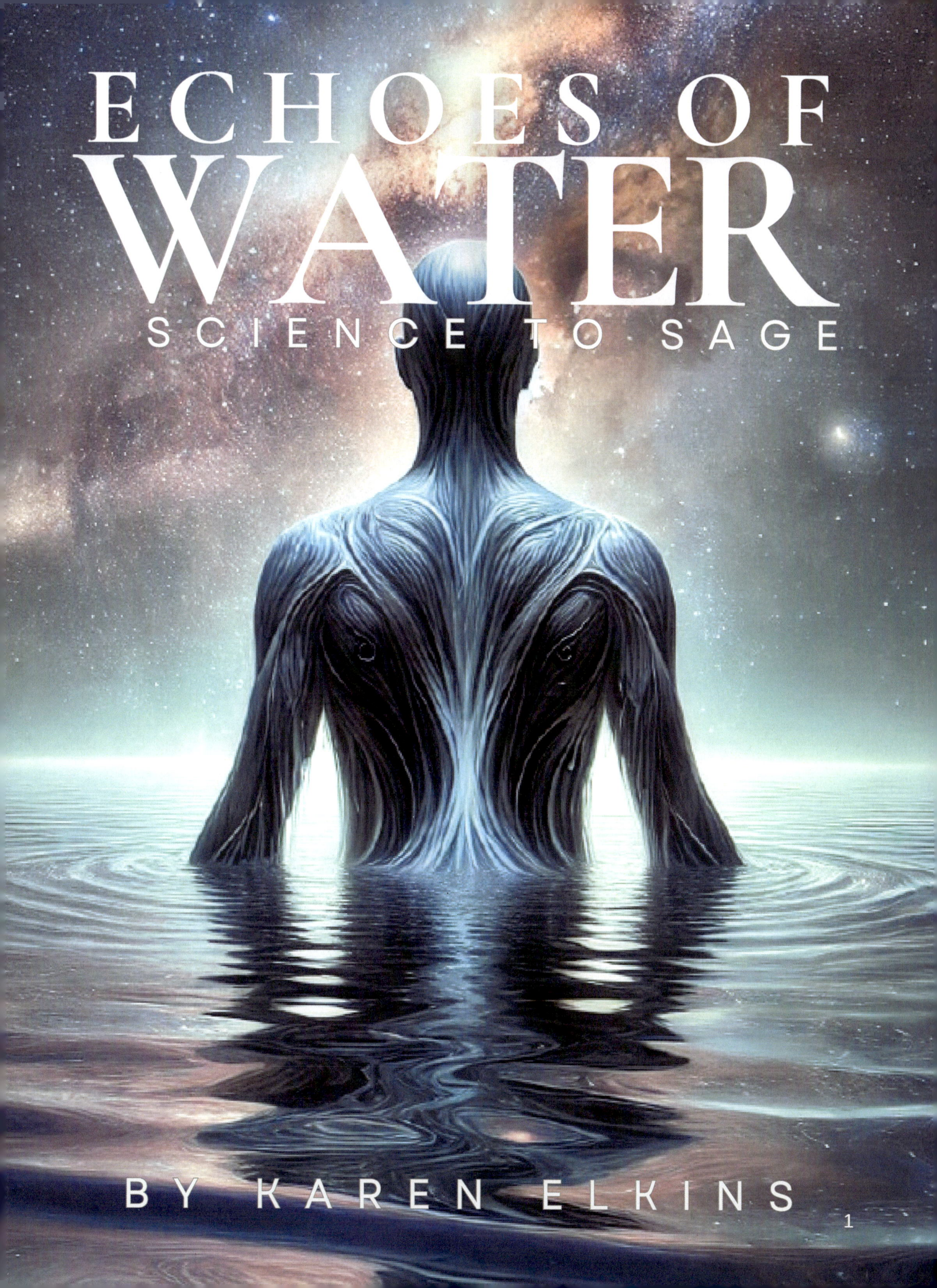

ECHOES OF WATER

SCIENCE TO SAGE

BY KAREN ELKINS

A PERSONAL INVITATION TO THE WONDERS OF WATER

Introduction

EXPLORE ANCIENT MYSTERIES AND THE MODERN SCIENTIFIC DISCOVERIES OF WATER

Welcome to a journey beyond the ordinary, an immersion into the boundless matrix of water, where science and mysticism merge to reveal the mysteries of life. Here, water is not merely a substance, but a living medium—a messenger between worlds, an artist shaping forms, a symphony of vibrations connecting every facet of creation. This journey beckons you to witness water's role as the conductor of cosmic patterns, the architect of nature's profound harmony, and the silent keeper of ancient memory.

As we delve into this realm, water's fluid grace unveils its intricate design. In crystalline lattice structures and whirling vortexes, water weaves the language of sacred geometry, channeling life's energies and balancing opposites. From its primal resonance in ancient oceans to its transformative power in cells and currents, water's alchemical essence is alive, disassociating impurities, cleansing, and renewing. Like a cosmic dance, each wave, drop, and ripple reflects an elegant rhythm, a universal harmony in which life's mysteries are encoded.

In water, we find the story of our own existence. We see how each molecule, from the smallest rivulet to the mightiest river, mirrors the same patterns that shape our very cells. This is the living matrix through which life flows, where words resonate and thoughts imprint. And just as water sustains the land, so it holds our collective memories, responding to light, sound, and energy in a timeless symphony that is as ancient as it is alive.

Prepare to traverse both scientific landscapes and mystical depths as we explore water's role as the bridge between the seen and unseen. Here, where science reveals its most profound insights and mysticism uncovers its deepest wisdom, we encounter water as a universal architect–a form that flows yet holds, that dissolves yet creates, a conductor of coherence, guiding us through life's unfolding mysteries.

This journey invites you to listen to the song of water, to see it as the quiet historian and vibrational artist that connects every wave of being. Step into the depths of this remarkable matrix, where the forces of creation whisper and the mysteries of the universe come alive, awaiting your discovery in each drop.

"Water flows not only to join rivers and seas but to bridge worlds seen and unseen, carrying life's whispers between shores."

CHAPTER 1

THE ETERNAL ELEMENT

FROM CHAOS TO CREATION

THE BIRTH OF THE UNIVERSE

In the earliest visions of the cosmos, ancient philosophers and mystics like Anaximander proposed that the universe arose from a primal unity, a substance so vast and boundless that it was beyond definition. He called this substance apeiron–the unlimited, a divine and perpetual essence, less concrete than the familiar elements of water, air, fire, and earth. This idea of apeiron echoes the notion of a cosmic egg, an undifferentiated mass holding the potential for all existence, where all elements were in a formless state.

This primal chaos–a vast, formless expanse filled with potential but devoid of structure–was the birthplace of the universe. It was a nebula of infinite possibility, waiting to be shaped into form. This primordial soup was the matrix from which all matter and energy would emerge. It contained the raw building blocks of life, suspended in the vast waters of creation.

As ancient myths and biblical texts describe, the act of creation began when the firmament—an invisible dome separating the heavens from the earth—was brought into being. From this moment, the chaotic waters were divided, creating order from chaos: *"And God said, 'Let there be a firmament in the midst of the waters, and let it divide the waters from the waters.' And God made the firmament, and divided the waters which were under the firmament from the waters which were above the firmament: and it was so. And God called the firmament Heaven."* - Genesis 1:6-8

This firmament represents the boundary between heaven and earth, the point at which structure and life began to take form. It is the protective barrier that holds the cosmic waters, and its gases condense to form the denser matter of Earth, allowing creation to crystallize into a more solid state, slowing the experience of "time."

In ancient Egyptian, Sumerian, and Babylonian myths, the idea of the primordial river flows as a central motif. This river symbolizes the lifeblood of creation, spiraling outward from the divine, its ripples giving birth to the world and all that exists within it. Each wave of this river represents a ripple of existence, crystallizing into form and life, much like the primordial soup evolving into complex organisms.

This echoes the idea that the divine crafts the structure of the universe, walking upon the waves of chaos and breathing order into the universe.

Through this lens, we see the interplay between chaos, the firmament, and the primordial soup as foundational to creation. It reminds us that from disorder, beauty, structure, and life emerge, weaving together the tapestry of the cosmos. The firmament serves as both a boundary and a protector, dividing the waters, allowing life to emerge, and shaping the very fabric of existence.

In the beginning, there was only Chaos, vast and dark, from which Gaia, Earth, came forth and then the wide waters of the Sea." - Hesiod's Theogony

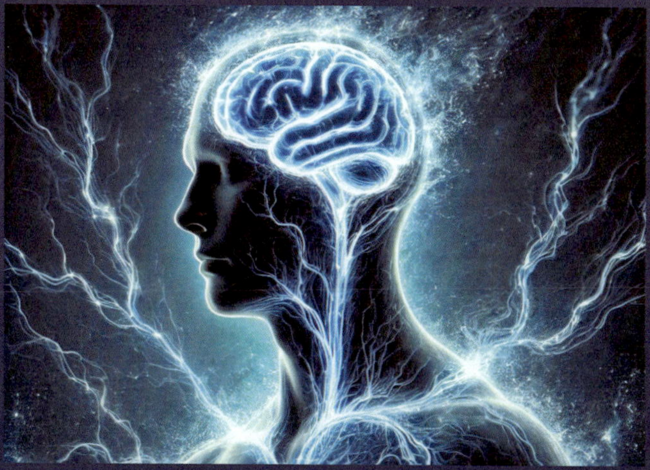

As Job 9:8 says: *"He alone stretches out the heavens and treads on the waves of the sea."*

"Variance in dimension of a substance does not change the character of the substance; it but changes state." ~ Walter Russell

GENESIS

CREATION AND THE PRIMORDIAL WATERS

In Genesis, *it says, "Let there be Light,"*

Brahma speaks *"aham, I AM, creating the wave, the optical illusion and vibrating the universe into existence."*

LANKAVATARA SUTRA

WATER IS THE BOUNDLESS OCEANS'

Universal Mind exists like a vast and boundless ocean. Waves disturb the surface, but beneath, all is calm and eternally unmoved. Having no personality, all things exist in it. But due to the disturbance upon its surface, it became an actor playing many parts." - The Lankavatara Sutra "

The Dogon people of Mali possess a cosmology as vast and intricate as the stars they revere, where water is seen as both a life-giving force and a cosmic connection. For the Dogon, water is the original essence, the primordial liquid from which the universe was born, stirred by Amma, their creator. This belief is woven into every layer of their cultural life and spiritual practices, where water, cosmos, and humanity reflect one another in cycles of creation and renewal.

DOGAN COSMIC BRIDGE

THE COSMIC EGG

As In Dogon myth, the universe began from a cosmic egg, embodying the potential of all life within its sacred waters. From this egg emerged the Nommo, revered as spiritual beings sent by Amma to bring order and wisdom to creation. The Nommo, deeply associated with water, are often depicted as amphibious, bridging worlds in their role as protectors of knowledge and bearers of the sacred. Through their teachings, the Dogon were gifted an understanding of water's life-giving rhythms:

"The Nommo taught us the secrets of water, the river that flows through the world and the stars alike. In their presence, we learned that water speaks in rhythms and patterns, giving us a language to touch the divine."

The Dogon people believe that, as we observe the flow of water in rivers and rains, we glimpse the same rhythms present in the stars. Sirius, the "eye of the heavens," holds a sacred place in Dogon cosmology, representing cosmic order and the balance of energies that water embodies.

Their knowledge of Sirius B, a star unseen to the naked eye, speaks to the depths of their astronomical and mystical insight, with water serving as a medium that connects earthly and celestial realms.

"Sirius is the eye of the heavens, watching the flow of water through the cosmos and guiding our souls along the river of life."

For the Dogon, all life is interconnected, much like the current that unites oceans and rivers. Each drop and each star play a part in the grand rhythm of creation, reflecting the unity that Amma instilled in every aspect of existence.

"All is one in Amma's creation; from water to earth, from earth to sky, the rhythm flows through all."

Thus, water in Dogon philosophy is more than a physical element–it is a cosmic bridge, a guide, and a sacred reminder of the interconnectedness of life and the wisdom within each wave, ripple, and drop.

BHAGAVAD GITA

ENERGY OF DESIRE

"I am the taste of water, the light of the sun and the moon..."
— Bhagavad Gita, Chapter 7, Verse 8

"As the rivers flow into the ocean, but the ocean is never filled, so flows the energy of desire into the mind of the seer, but they are never disturbed."
— Bhagavad Gita, Chapter 2, Verse 70

HONORING SACRED WATER &SPIRIT

STEWARDSHIP OF OUR LIVING WEB

Native wisdom teaches that water is a living being with spirit, memory, and purpose, deserving of reverence and care. Many Indigenous traditions view water as not just a resource but a sacred presence, a guide through the landscapes of both Earth and our inner lives. The Anishinaabe people refer to water as "Nibi," a word that conveys water's role as a conscious and connected spirit. This understanding of water as alive encourages rituals and ceremonies that honor it, recognizing that water can sense our intentions and respond to the care or disregard we show it.

The image of the dream catcher serves as a powerful reminder of this journey, symbolizing the dream of harmony between humanity and nature. Just as a dream catcher filters good intentions, allowing positive energy to flow freely while capturing negative influences, our journey with water invites us to filter our actions, keeping what sustains life and discarding what causes harm. Our role as stewards becomes a path of honoring the natural rhythms, cycles, and purity of water, acknowledging that as we care for water, we care for ourselves.

Through ceremonies, prayers, and acts of protection, Indigenous water protectors remind us of our shared responsibility to preserve this life-giving force for future generations. They stand as guardians, affirming that "Water is Life," a teaching that extends beyond individual belief systems into a universal call for collective stewardship. Protecting water is more than a duty–it is an acknowledgment of the sacred relationship between humanity and nature, one that transcends borders and unites us across generations.

As we explore this journey, may we carry forward the teachings of reciprocity and interconnectedness. May we embrace the wisdom of the dream catcher, understanding that our choices impact the entire web of life. To honor water as sacred is to recognize that its spirit flows within us, reminding us that our journey is one with the lands and waters we inhabit. By living as stewards, we nurture a path of respect and harmony, leaving a legacy that future generations can carry forward in reverence for this beautiful, interconnected world.

“The Great Spirit is in all things: he is in the air we breathe. The Great Spirit is our father, but the Earth is our mother. She nourishes us; that which we put into the ground she returns to us.” — Big Thunder (Bedagi), Wabanaki Algonquin

“Treat the earth well: it was not given to you by your parents, it was loaned to you by your children. We do not inherit the Earth from our ancestors, we borrow it from our children.” — Ancient Indigenous Proverb

Only when the last tree has died and the last river been poisoned and the last fish been caught will we realize we cannot eat money.” — Cree Prophecy

"The sound of water says what I think." — Chuang Tzu

THE TAO TE CHING

WATER AS THE WAY OF HARMONY

"Water flows in the way of the Tao, knowing when to yield and when to erode."

"Nothing in the world is as soft and yielding as water. Yet for dissolving the hard and inflexible, nothing can surpass it."— Lao Tzu

The Māori people of New Zealand hold water, or wai, as a sacred entity, embodying ancestral lineage and the essence of life itself. In their worldview, water is not merely a physical resource; it is a taonga (treasure), with its own spirit and mana (spiritual power), deeply intertwined with the essence of the land and sky. Every river, lake, and ocean embodies whakapapa–a genealogy linking the Māori to their ancestors and the gods.

MĀORI HONORING WATER'S LIFE-GIVING

Central to the Māori belief system is Tangaroa, the god of the sea, regarded as the source of all life. As guardian of the waters, Tangaroa represents both creation and continuity, with each wave carrying the history of those who came before. Māori prayers and songs are offered to Tangaroa, honoring water's life-giving powers and expressing gratitude for its role in sustaining communities.

"Our waters are sacred, a gift from Tangaroa. As they flow, they bring life to all that grows, all that breathes, and all that remembers."

The concept of mauri is key in Māori philosophy. Mauri is the life force or vital essence within every natural entity. Rivers and lakes are seen as beings with mauri, carrying the power to nurture or heal. Māori ceremonies often involve blessing water sources to restore or protect their mauri, recognizing that water's vitality directly affects the wellbeing of people and the land.

Mauri flows with every river; in its currents, we find ourselves, our ancestors, and our future.

To the Māori, water is also a means of connection. The rivers are seen as arteries linking the mountains to the sea, symbolizing the eternal flow between generations and the continuity of life.

"From the sacred springs to the ocean's embrace, water carries our stories, our ancestors, and our spirit."

Thus, for the Māori, water is not only a physical substance but a bridge to the divine, a keeper of history, and a symbol of unity.

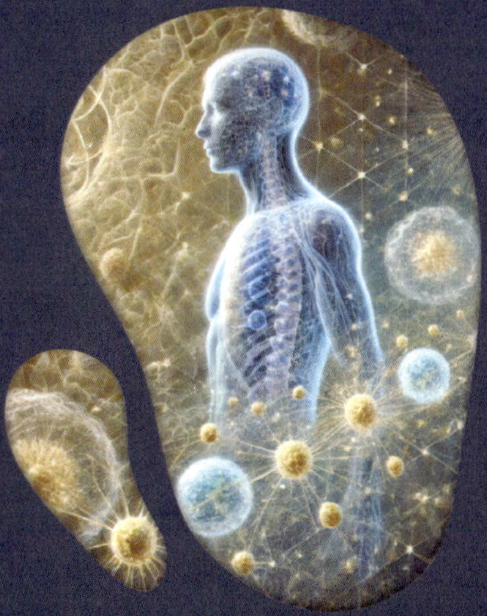

CHAPTER 2
WATER'S ROLE IN YOUR BODY

COSMIC TAPESTRY OF CREATION

INTERCONNECTEDNESS OF WATER, PARTICLES, AND LIGHT AS THE FUNDAMENTAL BUILDING BLOCKS OF CREATION

Matter is composed of three fundamental elements: water, particles, and light. Together, these components form the building blocks of existence, giving rise to the intricate structures and phenomena we observe in the universe.

As Khalil Gibran beautifully captures: *"In one drop of water are found all the secrets of all the oceans; in one aspect of you are found all the aspects of existence."*

Each drop of water, seemingly simple, holds the essence of the vast oceans and is made of both oxygen and hydrogen–elements that symbolize completeness and life itself. The landscape of water mirrors the cosmos, illustrating the vastness and depth of creation contained within each individual and each particle.

Water, in its transformative capacity, serves as a medium for life, facilitating interactions and reactions that are essential for existence. Particles, whether they are atoms or subatomic components, provide the substance and structure necessary for matter. Light, the energy that animates these particles, drives processes that sustain life, from photosynthesis to cellular respiration.

The difference between water in various states–liquid, solid, and gas–lies primarily in the density of the particles. This variation underscores the dynamic nature of matter, which never truly dies but instead recycles through different forms. Just as water travels through the cycles of nature, matter transforms and persists in new configurations.

As we explore the cosmos and the microcosm within ourselves, we uncover the truth that each aspect of existence is a reflection of the whole. In recognizing that we are composed of the same fundamental elements that make up the universe, we appreciate our place within this magnificent tapestry of life, ever woven by the forces of nature.

In the words of Albert Szent-Györgyi:*"Don't you realize that the sea is the home of water? All water is off on a journey unless it's in the sea, and it's homesick, and bound to make its way home someday."*

This reminds us of water's journey, echoing the cyclical nature of matter and its profound connection to life itself.

"There is a reciprocal (rhythmic balanced interchange) nature of the dynamical universe and correlates ancient earth, water, air and fire with modern chemistry and physics." ~ Dale Pond, Sympathetic Vibratory Physics

"Like water, we journey through cycles of stillness and movement, rising to new heights and returning to our depths. In each transformation, we carry the wisdom of every drop, every ripple—a quiet reminder that we are both the wave and the ocean."

TRANSFORMATION OF THE HUMAN BODY THROUGH STATES OF MATTER:

A JOURNEY OF BECOMING

"Arrival as Particles: The Dance of Wave-Particle Duality"

"From the cosmic fabric, a being emerges as light and particles, vibrating at frequencies beyond human comprehension. This is the quantum state of becoming, where potential and manifestation coexist."

- **Science:** Inspired by quantum entanglement, particles form a holographic projection—every part reflecting the whole.
- **Spirituality:** This phase represents the soul's descent from infinity into the finite, carrying the divine spark of creation.
- **Water's Role:** Water, the medium of life, exists in its vaporous form here, linking particles and creating a field for manifestation.

"Becoming Solid: The Crystallization of Energy"

"As vibration slows, particles crystallize into the structure of matter, forming a living lattice of crystalline water and biomolecules. The being takes shape, embodying the sacred geometry of the cosmos."

- **Science:** This mirrors the crystallization process, akin to water forming hexagonal ice lattices, stabilizing energy into organized matter.
- **Spirituality:** The soul becomes flesh, fulfilling its divine blueprint as a living temple for spirit.
- **Water's Role:** The body forms through water's crystalline properties, structuring and transmitting energy in perfect harmony with the universal design.

"The State of Crystalline Existence: Light Embodied in Matter"

"Fully formed, the human becomes a harmonious interplay of light and matter, where the crystalline structure of cells channels energy and consciousness."

- **Science:** Biophotons emitted by DNA allow cellular communication through light, creating a coherent field of energy.
- **Spirituality:** This represents spiritual maturity–a crystalline state of clarity, purity, and alignment with universal flow.
- **Water's Role:** Water acts as the matrix of life, a conductor for light and consciousness, amplifying the interplay between physical and spiritual realms.

"Evaporation: Dissolving Back into Energy"

"At the end of the physical cycle, the body releases its form, ascending into less dense states. Water evaporates, and the being merges into the infinite field of energy."

- **Science:** Like sublimation, where solids transform directly into gases, this phase symbolizes a direct return to an energetic state.
- **Spirituality:** The soul is liberated, returning to its infinite source–a rebirth where consciousness merges with universal flow.
- **Water's Role:** The evaporation of water mirrors the ascension of the soul, releasing its physical bonds and dissolving into the infinite.

"The Cycle of Return: A Fractal Existence"

"Essence lingers as a fractal in the cosmos, ready to reincarnate. Each transformation is a continuation, embodying eternal change and growth."

- **Science:** The conservation of energy states that energy cannot be created or destroyed, only transformed.
- **Spirituality:** This reflects the infinite dance of creation, where every end births a new beginning.
- **Water's Role:** Water's cycles of evaporation, condensation, and crystallization serve as the perfect metaphor for the eternal journey of the soul.

ALBERT SZENT-GYÖRGYI

WATER AS THE MATRIX OF LIFE

Albert Szent-Györgyi, a Nobel laureate in Physiology, viewed water as the essential element of life. He famously described water as "the matrix of life," emphasizing its role in cellular processes, including photosynthesis and energy transfer. In Szent-Györgyi's view, the sea embodies life's origins, containing the salts, minerals, and energies necessary to support biological processes.

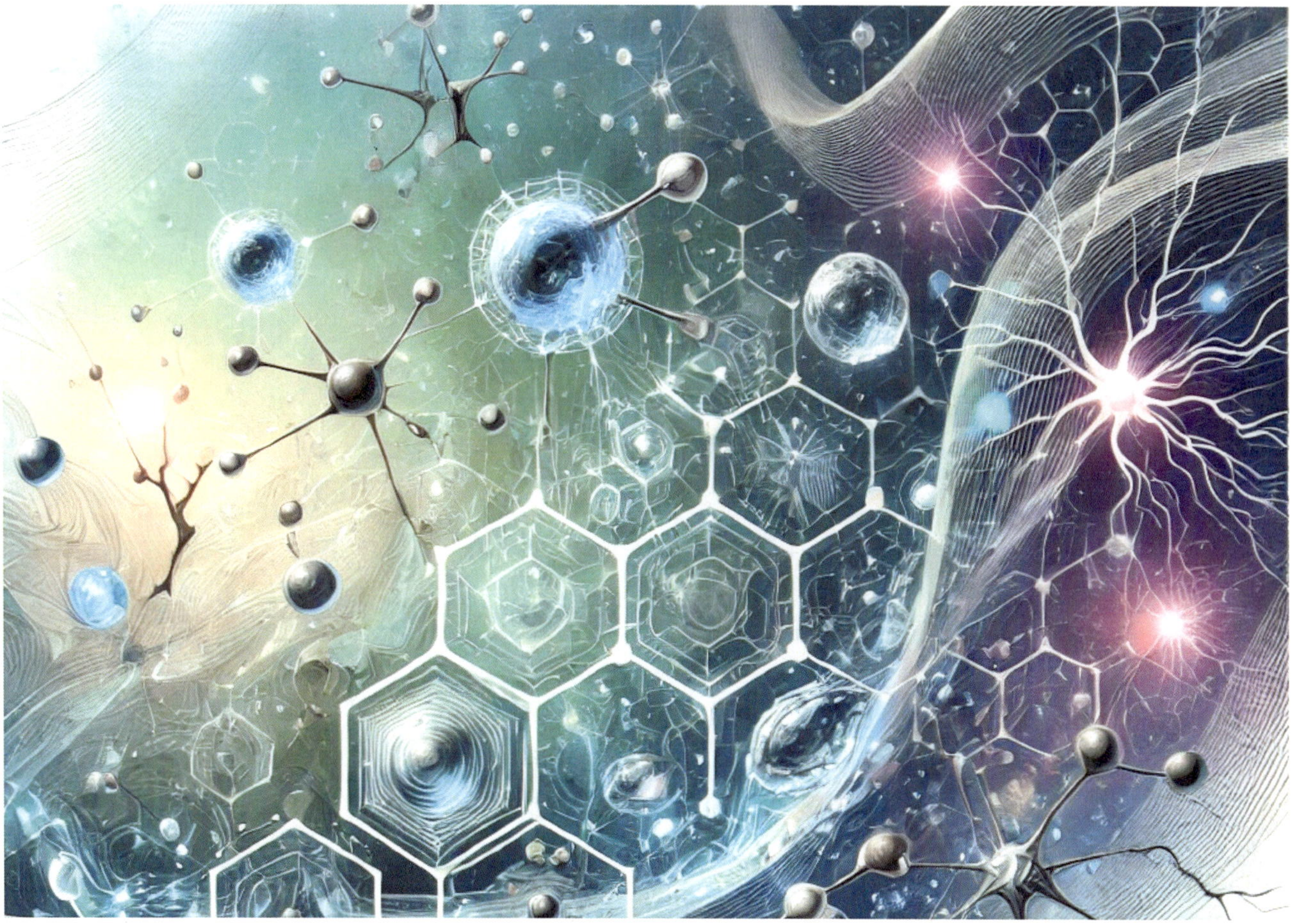

Szent-Györgyi saw water as fundamental to life's architecture. To him, water was not just a supporting player in biological processes but the essential framework within which life occurs. He believed that life depends entirely on water's unique properties, making it the "matrix" that nurtures and sustains every cell.

"Life originated in water, just as every cell is filled with it. To understand water is to understand life itself."

For Szent-Györgyi, the history of life on Earth is a story told through water. He believed that by studying water's molecular structure and behavior, we could unlock the secrets of cellular health, energy flow, and even the origins of consciousness *"I owe my health to water, and if life has an essence, it flows within every drop."*

This quote captures Szent-Györgyi's view that water doesn't merely carry life–it embodies it. In his perspective, each molecule of water contains a spark of life, connecting all organisms in a shared biological tapestry.

"Water is life's matter and matrix, mother and medium. There is no life without water."

Szent-Györgyi's work emphasized that water is central to biochemical reactions, cellular respiration, and the transport of nutrients. He recognized that water's molecular bonds, structure, and flexibility make it a perfect medium for supporting complex life. His insights help frame water as the essential life force, a substance that bridges the physical and the biological, sustaining life at every level.

"Water has a memory and carries within it our thoughts and prayers. As you will see, this truth is far beyond any understanding of modern science." - Dr. Masaru Emoto

"You are not a drop in the ocean. You are the entire ocean in a drop."- Rumi

"The material body consists of fire, water, earth and air, this is our mortal and earthbound nature. Our Soul is nourished by fire and air, and our body by water and earth." ~ The Wisdom of the Pharaohs

WATER BODIES

THE ALCHEMY OF WATER AND THE ELEMENTS ROLL

At a molecular level, the human body is composed of approximately 99.98% water molecules. The brain, in particular, holds an astounding 90% water content, while the body as a whole typically ranges between 60-70%, depending on factors like age and hydration. This profound water composition plays a pivotal role in our physical and cognitive functions.

Consider the notion that the brain, with its high water content, serves not only as the control center for bodily processes but also as a transformative medium for consciousness. Water's unique ability to respond to energetic and vibrational changes aligns seamlessly with the concept that our thoughts create ripples within us, "transforming" the water at a cellular level and influencing our perception of reality. This responsiveness positions the body as a vessel of liquid alchemy, where air, fire, and water regulate and intertwine, placing us in a perpetual state of chemistry in motion. These elements reflect and amplify our fluctuating emotions and shifting mental states.

This interconnected system within the body functions as a dynamic network, enabling the transmission of electrical signals, communication between cells, and overall coordination of bodily functions. The living matrix comprises proteins, water, and other biomolecules that form an integrated support structure, facilitating the seamless exchange of information and energy.

As Jim Oschman, Ph.D., states in Energy Medicine, The Scientific Basis: "*... In essence, our bodies are composed mainly of materials that are similar in properties to the substances that make possible our modern computers, cell phones, televisions and so on. The living matrix concept reveals the cell and its environment as composed of a continuously interconnected system.*"

As the mind embraces new thoughts, the water in the brain acts as an alchemical medium, recording and mirroring these changes. This dynamic process allows us to experience life through an altered, elevated state of awareness. Much like alchemy, water becomes an agent of transformation within us, helping to "dissolve" old patterns and foster the embodiment of new ones. This continuous transformation highlights how water supports not just our physical well-being but also our journey of emotional and spiritual evolution.

WOMB OF CREATION

WATER IS THE WOMB OF CREATION FROM WHICH ALL IS BIRTHED

We are born within our mother's womb, embraced by water–our first ocean, a cradle that mirrors the ancient seas from which all life once emerged. In this sacred space, we develop within a cocoon, wrapped in the elegance of the golden ratio, an eddy-like spiral guiding our growth. This divine pattern, like a gentle tornado, spins through creation, directing the unfolding of life with the precision and grace of nature's design.

As we emerge into the world, we carry with us the rhythm of the ocean, the harmony of the universe's design, and the wisdom of nature's code. Each heartbeat and each breath echoes the primordial waters, reminding us of life's endless dance in alignment with creation itself. The golden spiral, the flow of water, and the pulse of life are intertwined, shaping us as part of the infinite tapestry of existence.

"In every drop of seawater lies the memory of life's origins, an eternal blueprint that nourishes, heals, and sustains." - Rene Quinton

"The very water we are conceived in the womb is hexagonal in structure, both inside and outside of cells."
~ Gerald Pollack, Ph.D.

RENE QUINTON

THE HEALING POWER OF SEA AND NATURE'S HIDDEN SECRETS

Rene Quinton (1866–1925) was a pioneering French biologist and physiologist who dedicated his life to understanding the origins of life and the remarkable parallels between seawater and human biology. Fascinated by the evolutionary connection between marine environments and human physiology, Quinton observed that seawater closely mirrors the composition of human blood plasma, suggesting a deep biological link to our oceanic beginnings. This insight led him to develop "Quinton's Plasma," an isotonic solution of purified seawater designed to restore cellular health. Quinton believed that, in essence, the ocean's water was a "living plasma" that mirrored our own internal environment.

"Seawater is not merely a source of nourishment, but a living bridge, harmonizing with the body's cells and connecting us to the ancient waters from which life first arose."

In groundbreaking experiments, Quinton used his Plasma solution to replace blood, demonstrating seawater's compatibility with bodily fluids and its restorative properties. His work highlighted the idea that we carry a trace of the ocean within us—a "memory" of our origins embedded in the very composition of our blood. Quinton's Plasma became a therapeutic tool for revitalizing cells and enhancing vitality, and it was widely used in Europe for various health applications.

Through his research, Quinton revealed a profound unity in life's origins, suggesting that our internal ocean sustains us in the same way the primordial sea sustained early life. His work illustrates how the very elements that make up seawater—sodium, chloride, potassium, calcium, and magnesium—are essential to human life, linking us to that ancient source of vitality. Quinton's legacy endures in the understanding that we are, in essence, vessels of the ocean's life-sustaining rhythm, carrying a timeless connection within us to the waters that first nurtured life on earth.

"Our blood is but seawater in another form, a continuation of the life force that began in the ocean."

René Quinton devoted his life to unveiling the profound healing powers of seawater, ultimately introducing the world to Quinton Marine Plasma. At the turn of the 20th century, his positioned seawater as more than a nutrient source—it was a sacred, life-sustaining medium. His discoveries led to the development of Quinton Marine Plasma, a specially prepared and purified solution of seawater intended for therapeutic use. Through numerous clinical trials, Quinton applied this plasma to treat patients with immune deficiencies, skin disorders, and digestive and respiratory issues, achieving remarkable success and solidifying his legacy as a visionary in natural healing.

THE SECRET VARIATIONS OF WATER

LIGHT, MEDIUM, AND HEAVY IN THE HUMAN BODY

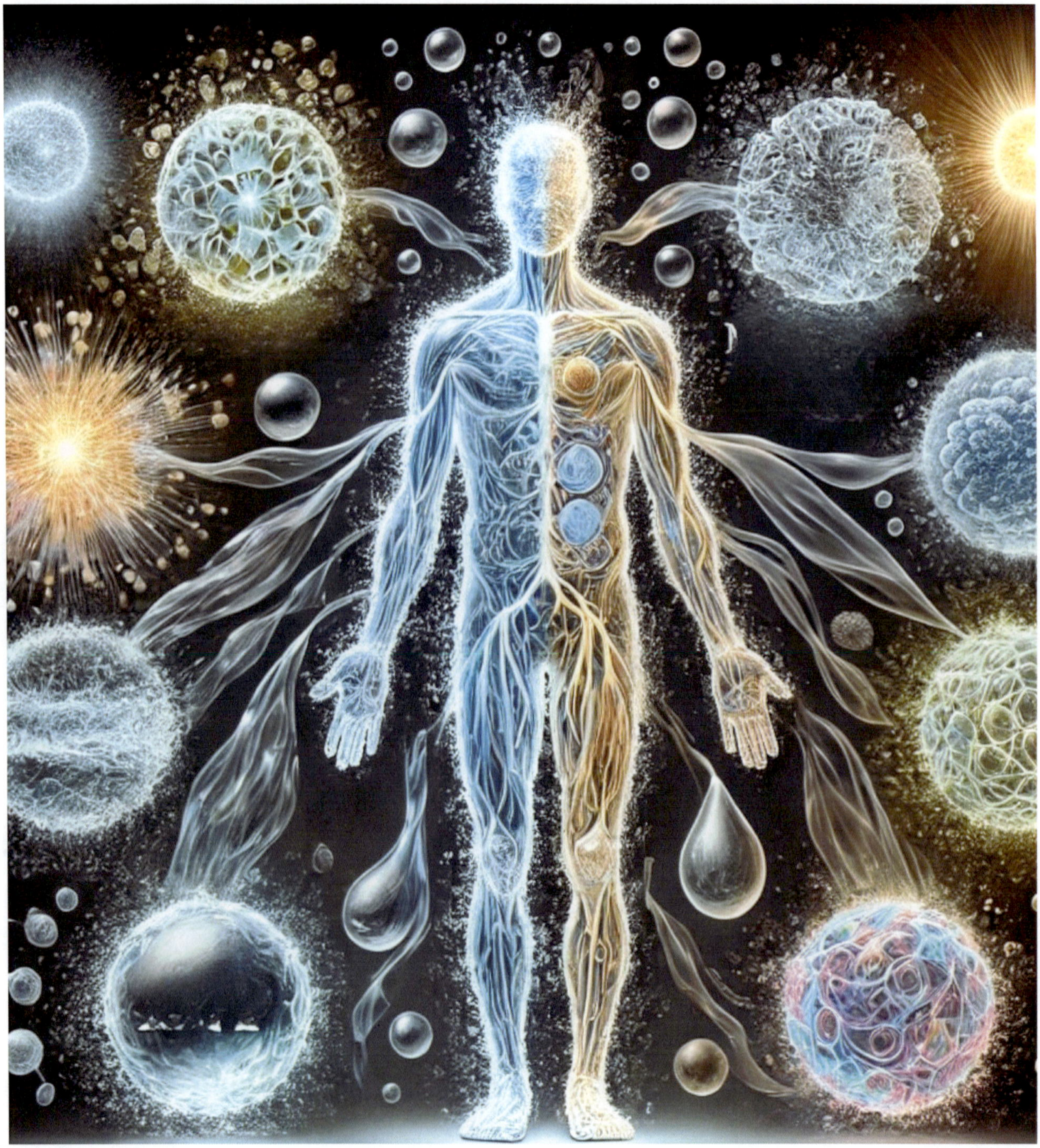

Water exists as three distinct forms—light, medium, and heavy—each possessing unique densities and properties that embody the delicate balance between creation and transformation. This triad of forms creates a spectrum where water's molecules, like dancers, shift and flow, mirroring the dual essence of Shiva. Light water moves gently, seamlessly nourishing and supporting the cycle of life, while medium water subtly affects energy exchange, resonating as a bridge between balance and change. Heavy water, dense with potential, holds within it the capacity to disrupt and transform, embodying the potent duality of existence as both a nurturer and a powerful force of alteration.

Light water, the most abundant form, nourishes and sustains all life, seamlessly supporting cellular energy production (ATP) and the countless biochemical reactions essential for growth, repair, and regeneration. It harmonizes with our natural rhythms, a gentle creator that fuels vitality within us.

Medium water, with its slightly altered properties, resonates differently in the biological realm, subtly shifting the energy exchange within our cells. This form of water maintains life's balance while initiating a nuanced energy transfer that supports transformation, an element of change woven into our cellular framework.

Heavy water, however, reveals water's hidden potency. With an additional neutron in each hydrogen atom, it disrupts ATP production, altering the motor of life's energetic processes. Walter Russell first observed this unique atomic variation, noting that heavy water's frequency impacts cellular structure and energy cycles. In essence, heavy water functions as a quiet "bomb" in biological systems, capable of reshaping cellular function at the molecular level. Its effects are profound, demonstrating how even a minor shift in water's atomic composition can influence the flow of life energy, balance, and health.

Understanding these variations unveils a secret power within water's atomic architecture, revealing how it can act as both a gentle creator and a potent transformer. Heavy water, once a hidden presence, now stands recognized as an element that disrupts life's chemistry and energy. In the human body, water becomes not only a sustainer of life but a powerful medium that nurtures, balances, and–when altered–disrupts our existence at a cellular level. These variations invite us to explore water's profound influence on health, energy, and the intricate dance between life's essential forces.

In the dance of water's molecules, we see the essence of Shiva—the power to create and the potential to destroy. Light water nourishes and sustains life, while heavy water holds a quiet, potent force capable of disruption. Water is both the gentle creator and the fierce transformer, embodying the balance of life's cycles.

TRIPLE POINT: TRIPLE-POWERED SPIRIT

THERMODYNAMIC STATE - SOLID, LIQUID, AND GASES

The triple point of water is a unique thermodynamic state where water exists simultaneously as a solid, liquid, and gas. This state occurs under specific conditions of temperature and pressure, creating a balance among all three phases. In this convergence, we observe the essence of transformation–a natural "trinity of life" that unites water, air, and plasma. These three elemental states are not just aspects of water; they are foundational to life itself, forming an elemental cycle from which all living things emerge, evolve, and recycle.

In the depth of winter, I finally learned that within me there lay an invincible summer." — Albert Camus

This triple point embodies a profound cycle of energy and life force. It illustrates how diverse forces unite, becoming something greater than the sum of their parts, just as an idea moves from imagination to solid form. In its fluid phase, an idea flows freely, shaping itself through imagination. With passion–our inner "fire"–the idea gains energy and moves into a dynamic state. Finally, in its solid phase, the idea takes shape and becomes real.

This "trinity of life" reflects nature's cycles of adaptability and creation, showing that life is born from the union of distinct yet complementary elements. Water offers fluidity, air brings energy, and plasma–the electrified, high-energy state–catalyzes transformation. This dynamic is echoed within our own system, as elemental forces shape our eleMENTAL nature–our fundamental chemistry and energy flow.

Recognizing this elemental balance can lead to a powerful realization: we, too, have a "triple point," a moment of "ah-ha" where we can pause, take a breath, and choose the energy we bring to any situation. By becoming aware of the elemental energies that influence us, we can consciously engage in life's cycles, using our imagination, passion, and action in harmony with our inner chemistry.

"The most powerful weapon on earth is the human soul on fire." — Ferdinand Foch

The triple point of water reminds us that transformation is inherent in life. It's not merely a metaphor, but a lived experience of balance that we can apply to our own lives.

This trinity of faces captures the essence of the triple point and "trinity of life" within a single, unified figure, emphasizing the interconnectedness of the phases. Each face not only symbolizes an elemental state but also represents a mindset: imagination, passion, and completion. This image serves as a powerful representation of how we can harmonize and harness our elemental nature to bring ideas from fluidity through fiery transformation into solid realization. It also gives each phase a distinct attitude and energy, embodying a cycle that reflects the inner journey of thought to action.

"Vision without execution is just hallucination." — Thomas Edison

"An idea that is developed and put into action is more important than an idea that exists only as an idea." — Buddha

DR. GERALD POLLACK

THE FOURTH PHASE OF WATER: A GATEWAY TO CREATION

Dr. Gerald Pollack's groundbreaking research on the "fourth phase" of water reveals it as a highly organized, crystalline structure (gel), distinct from the familiar solid, liquid, and vapor states. In this phase, water responds to energy–particularly light–by forming structured layers of molecules. This organized state acts as a framework capable of storing energy and information, making it a vital element in the assembly of life's building blocks.

Crystalline Matrix as a Blueprint for Creation

Pollack's discovery positions the fourth phase of water as more than a passive medium–it is an active participant in the structuring of matter. This crystalline matrix creates a highly energetic and ordered environment, attracting and organizing molecules into the foundational forms necessary for life. Acting as a "formative force," this structured state enables differentiation–the process through which biological forms diverge and evolve into distinct features, a cornerstone of most ecosystems.

The Role of Water in Biological Differentiation

Water's crystalline matrix serves as a medium for the initial patterns of differentiation in life. This matrix not only facilitates the organization of cellular structures but also supports the development of specialized biological functions. It may underpin the divergence of reproductive roles and the emergence of biological polarities, offering a blueprint for the organization of male and female characteristics in complex life forms.

Light and Water: Energizing Creation

Pollack emphasizes the role of light as a driving force in the formation of structured water. Sunlight and infrared heat fuel the transformation of water into its fourth phase, turning it into an active participant in creation. This connection between water and light echoes ancient understandings of life as a harmonious interplay of these two essential elements. Through this interaction, structured water becomes the dynamic interface where the material and energetic worlds converge, shaping both simple and complex life forms.

Water as a Dynamic Power Source

Within the human body, Pollack reveals, structured water plays a pivotal role in cellular energy. Each cell acts as a tiny battery, with positive and negative charges separated by molecular structures. The energy stored within these batteries is powered by the exclusion zone (EZ) water, a gel-like form that holds a slight negative charge. By absorbing electromagnetic energy from the environment, including sunlight and infrared radiation, EZ water continuously recharges these cellular batteries, fueling life at its most fundamental level.

The Matrix of Life's Architecture

Water's fourth phase aligns with proteins and DNA, stabilizing their structures and enabling the biochemical processes critical for life. This structured water creates the scaffolding for cells and tissues, forming the intricate architecture found across all living organisms. In multicellular beings, the crystalline state of water ensures coherence, supporting communication, growth, differentiation, and reproduction. This matrix acts as the organizing force behind the complexity of life, ensuring that systems remain interconnected and functional.

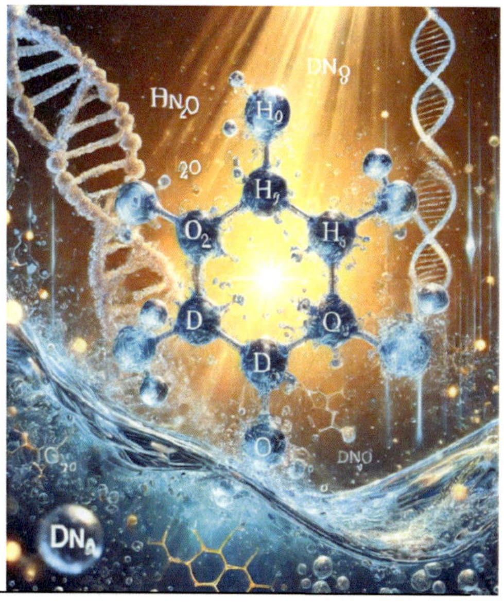

"The fourth phase of water is the bridge between biology and physics, uniting the energetic and material worlds in a single, dynamic form."

CELLULAR COMMUNICATIONS

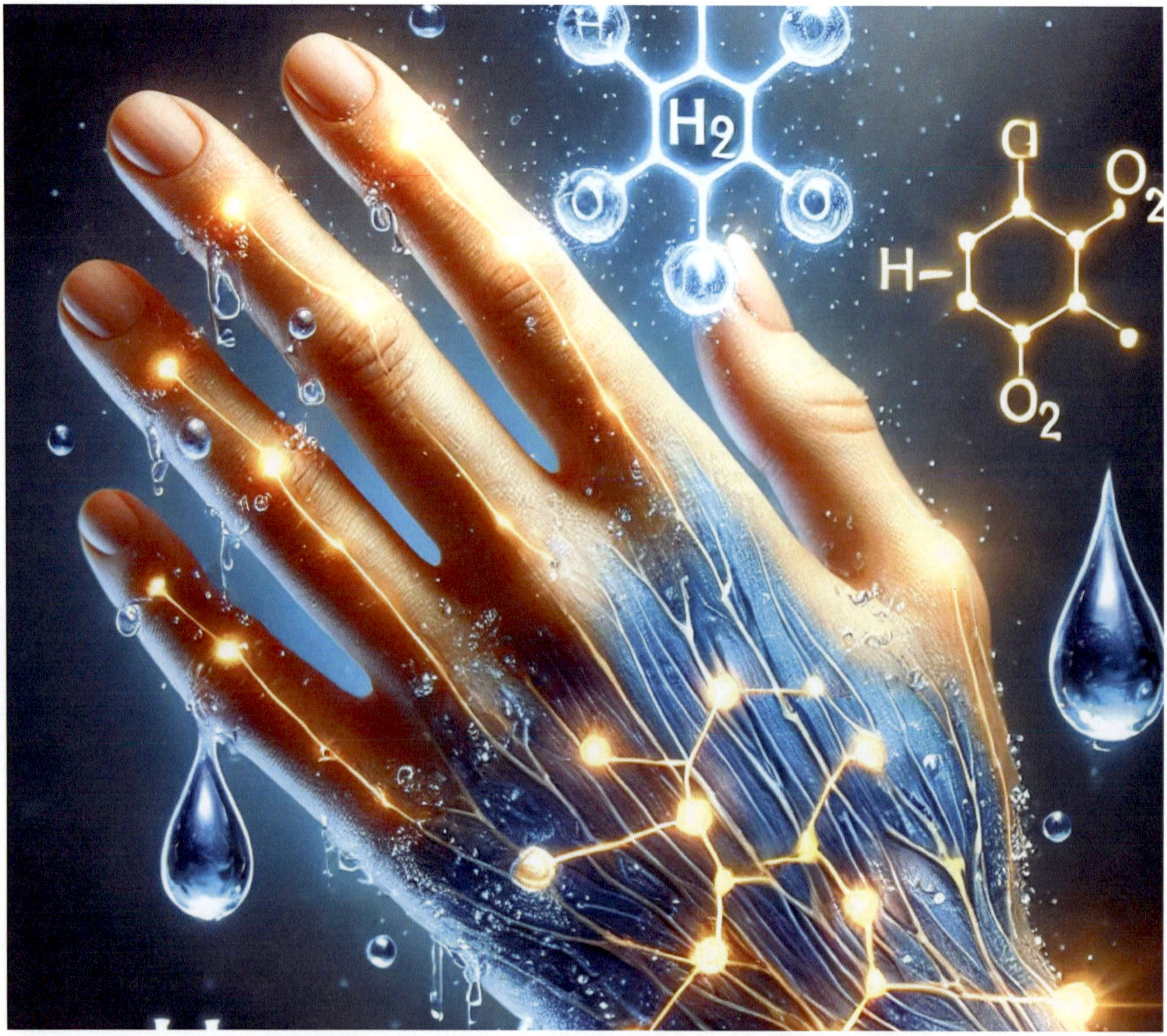

Bridging Biology and Physics

Pollack's insights suggest that the fourth phase of water is the bridge between biology and physics–a unifying force that connects the energetic and material realms. By acting as a bioenergetic medium, structured water enables the body to absorb and transform external energy into life-sustaining charge, underscoring its role as a dynamic foundation for creation.

From enabling cellular processes to organizing multicellular structures, water's crystalline matrix is the "sweet spot" from which creation emerges. It offers a profound insight into life's interconnectedness, illuminating the processes that allow life to differentiate and evolve. Pollack's work transforms our understanding of water from a simple solvent to a dynamic, structuring force that underpins the complexity and diversity of life on Earth. Through its partnership with light, water not only supports life–it shapes it, forming the foundation for the vast tapestry of existence.

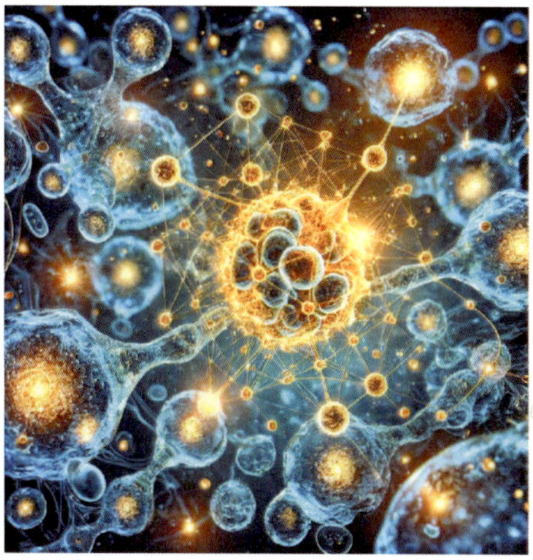

On a larger scale, water mirrors this function in nature, channeling energy through rivers, oceans, and clouds. Just as lightning strikes charge the atmosphere and feed the Earth's electric potential, water becomes the conduit that distributes this energy, carrying it into the soil, plants, and eventually into every living being. The flow of water is thus a carrier of life's electric charge, bridging the physical and energetic realms.

Water: Nature's Capacitor

Water's role as an electric bridge extends beyond its conductivity. Its unique structure allows it to interact with electromagnetic fields, amplifying and sustaining the charge that flows through the natural world. This is why water is so responsive to the environment, capable of transmitting not only energy but also the information embedded within it. Like a living battery, it holds the charge of life itself, continually replenishing and renewing the systems it touches. Whether in the human body, within cells, or in the rivers and oceans of the planet, water acts as the pulse of the electric web that connects all life. This electric flow binds the microcosm to the macrocosm, sustaining life through its dynamic, charged dance. Water is not just life's carrier–it is its power source, the endless current that unites and animates all living systems.

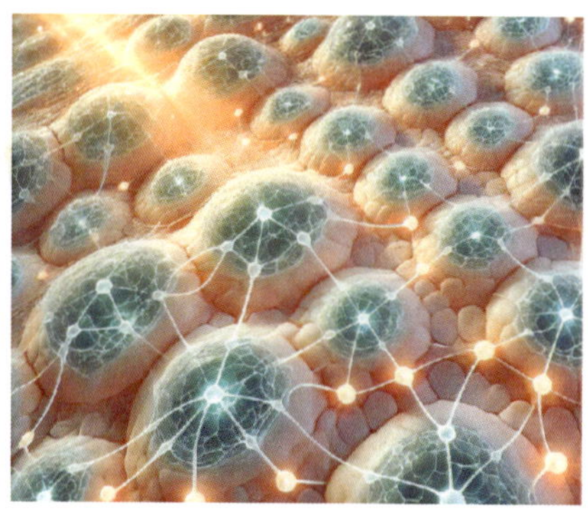

"Water, the lifeblood of the earth, flows within us, connecting our physical being with the pulse of the universe."

"Just as rivers carve through the land, water flows through our bodies, shaping life itself, nourishing every part of us."
– Anonymous

"As light electrifies water's perfect channels, it follows fractal trails that mirror the universe, showing life's energy held in every drop."

WATER'S POLARITY

DIPOLE STRUCTURE AND CHARGE SEPARATION

Water molecules have a dipole structure, meaning they have a positive charge on the hydrogen side and a negative charge on the oxygen side. When water molecules align in certain ways, they can create areas of positive and negative charge, much like the terminals of a battery. This alignment is especially prominent in structured or "ordered" water, such as water in biological cells or near certain surfaces, where molecules align to create distinct charge zones.

When exposed to certain conditions, such as light or certain minerals, water can form layers of positive and negative charges. This phenomenon, known as the Exclusion Zone (EZ) effect, was discovered by Dr. Gerald Pollack. In the EZ, water molecules organize themselves into a structured phase, separating into negatively charged zones and leaving protons (positively charged particles) nearby. This separation of charges creates a voltage potential—like a small battery—between the structured and bulk water.

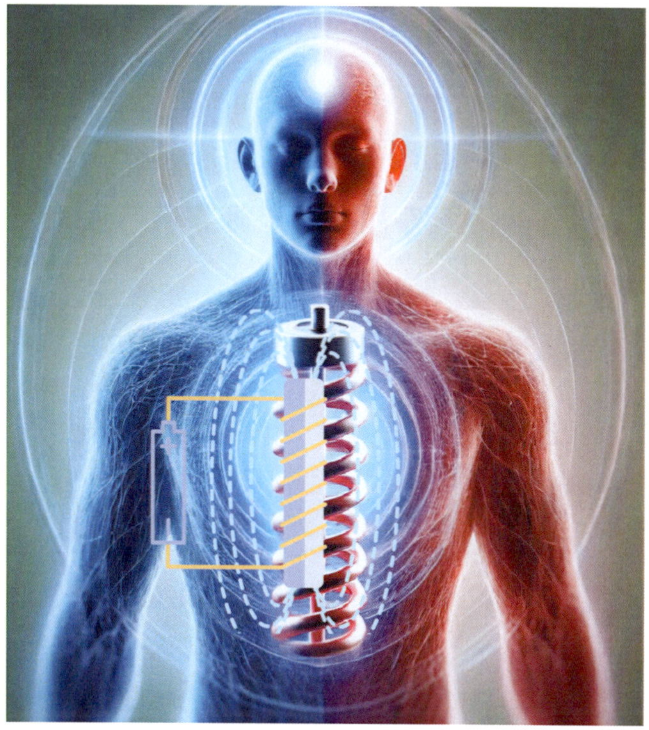

Water can absorb energy from external sources such as sunlight, sound, or other electromagnetic fields. This energy can excite the water molecules, causing them to separate charges further and enhance the EZ. The structured water can then store this absorbed energy and release it gradually, providing a consistent source of energy, similar to how a battery discharges. This stored energy is especially useful in biological systems, where cells use it to support cellular functions.

In biological systems, water acts as a medium for proton (H^+) exchange, which is essential in cellular energy processes. Proton flow through water channels within and around cells contributes to what is essentially an electrical current, allowing for energy transfer and communication across biological structures. This function is similar to the flow of electrons in a conventional battery.

Water's ability to act like a battery is enhanced by the presence of other elements, such as salts and minerals, which increase its conductivity. Ions (charged particles) from dissolved salts enable more efficient energy storage and transfer, much like an electrolyte in a battery.

In the human body, energy generation and "battery-like" function operate through biochemical processes rather than the electrochemical reactions in a typical battery. However, the principles of charge, ion movement, and electron flow are similar, especially in cellular energy production.

In summary, water acts like a battery by harnessing its dipole nature, organizing charge zones, absorbing energy, and facilitating proton flow—all of which enable it to store and release energy in a way that powers natural processes, particularly in living organisms. This ability to function as a bio-battery is one of the reasons water is so essential to life.

THE WATER BRIDGE EXPERIMENT

THE HUMAN BODY CONNECTION

Dr. van de Graaff's work with water bridges adds to the broader field of study concerning electrical conductivity in fluids, and it showcases the unique properties of water. His demonstrations are often used in educational contexts to illustrate concepts of electricity, fluid behavior, and the interconnectedness of physical principles.

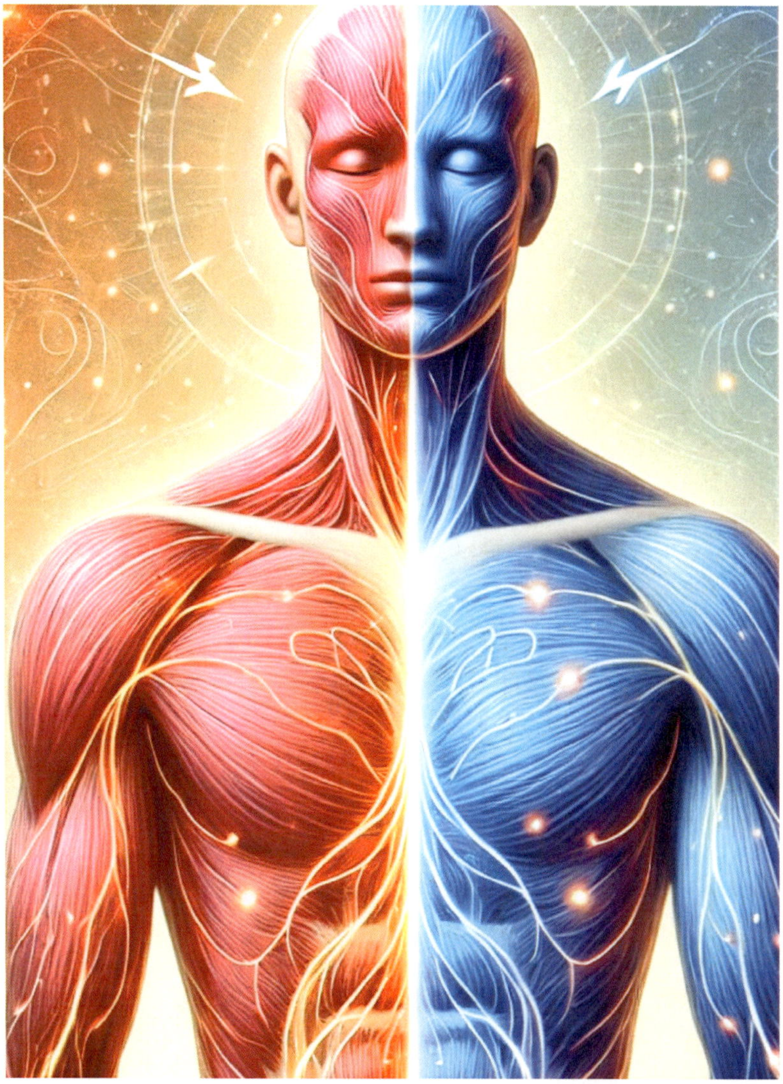

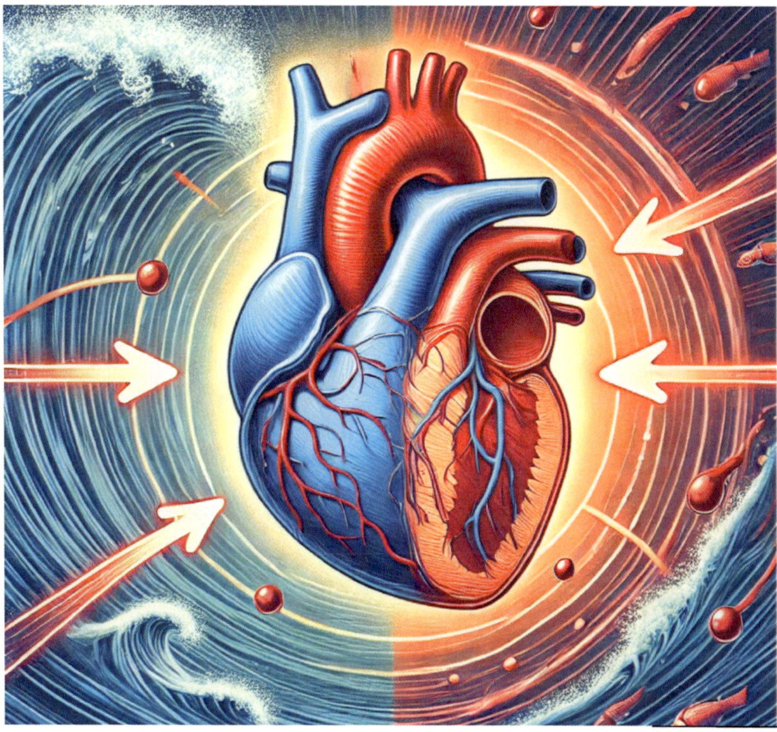

The concept of balance in the human body mirrors the principles observed in the water bridge experiment. In this experiment, two beakers of water are connected by a bridge of water, demonstrating how charge flows between two distinct areas. The electric charge seeks equilibrium, similar to how the body's two hemispheres and heart chambers work together to maintain homeostasis.

Just as the water bridge allows for the movement of charge between two points, the human body relies on the coordinated flow of energy and signals across its paired structures. The heart's chambers pump blood efficiently, while the brain's hemispheres communicate and integrate information, illustrating the body's innate design to achieve balance.

In essence, both the human body and the water bridge experiment exemplify the principle of duality and the quest for balance. The interplay between the two chambers of the heart and the flow of charge in the water bridge serves as a powerful reminder of how interconnected systems work to sustain life and harmony, whether in biological structures or physical phenomena.

The human body exemplifies remarkable symmetry and duality, characterized by its inherent structure. Central to this system is the heart, which comprises two distinct chambers: the left atrium and ventricle, and the right atrium and ventricle. Each chamber plays a vital role in the circulatory system, working in tandem to pump blood throughout the body.

WATER, THE LIFE BLOOD

BODIES - HUMAN TO OCEAN

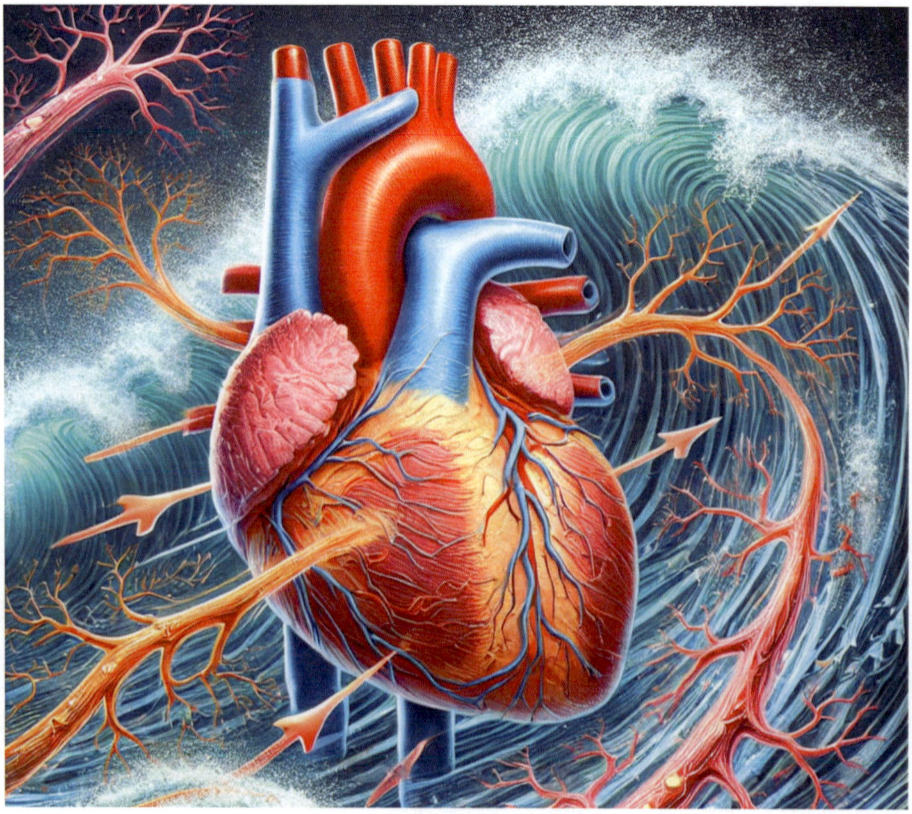

Water is often referred to as the essence of life, serving as the lifeblood for both the human body and the vast oceans that cover our planet. This remarkable substance plays a critical role in sustaining all forms of life, acting as a medium for essential biochemical processes and nourishing ecosystems.

The Heart of the Body

In the human body, water is crucial for maintaining homeostasis. It carries nutrients and oxygen to cells, removes waste products, and regulates temperature, all vital for health and vitality.

The heart, with its rhythmic pumping action, circulates blood—a fluid primarily composed of water—throughout the body. This circulation mirrors the flow of water, highlighting how interconnected we are with this life-giving substance.

Ocean Currents as Life's Rhythm

The currents of the ocean similarly reflect the flow of our heart's current. Just as the heart pumps blood, creating a rhythmic circulation system, ocean currents also act as pumps, moving water across vast distances. These currents transport heat, nutrients, and gases, supporting marine life and regulating the Earth's climate. This flow is not only physical; it represents energy in motion, embodying both charge and fluidity.

A Cycle of Connection

Both the human body and the oceans exemplify the essential role of water in creating balance and sustaining life. As we depend on water for our bodily functions, the oceans rely on their currents to maintain the health of marine environments. This cycle of connection reinforces the idea that water is not merely a resource but a vital force that links all living beings and systems.

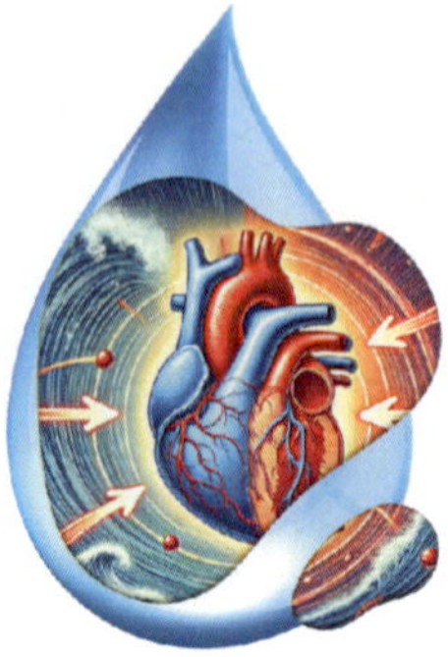

THE BRANCHING OF LIFE

VEINS OF EXISTENCE

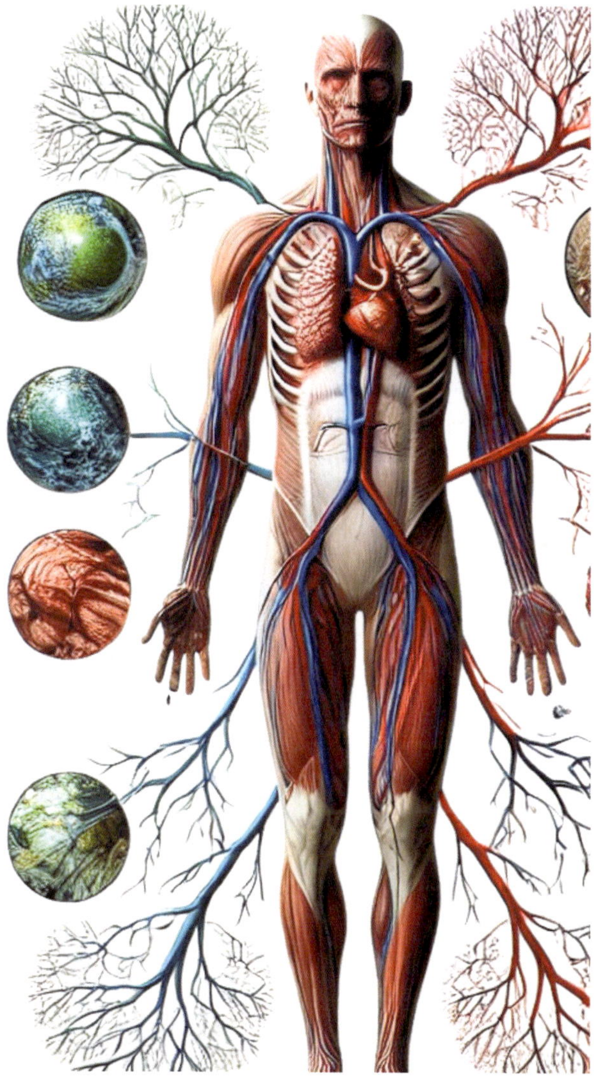

As water acts like a spark–a natural battery- it enables life to flourish. Again water's polarity allows it to act like a capacitor, holding an electric charge and releasing it to fuel cellular processes. This electrical potential flows like a river through our bodies, energizing every cell and sustaining the rhythms of life.

In the body, this electric river is most evident in the flow of ions, where charged particles are carried by water to maintain the balance of energy within and between cells. This constant movement creates a dynamic current, much like the flow of electricity in a circuit. The cytoskeleton within cells acts as a conductor, channeling these charges across the cellular network, ensuring energy flows seamlessly to support growth, repair, and communication.

On a larger scale, water mirrors this function in nature, channeling energy through rivers, oceans, and clouds. Just as lightning strikes charge the atmosphere and feed the Earth's electric potential, water becomes the conduit that distributes this energy, carrying it into the soil, plants, and eventually into every living being. The flow of water is thus a carrier of life's electric charge, bridging the physical and energetic realms.

THE SOLVENT OF THE UNIVERSE

WATER'S CLEANSING POWER

Water's cleansing power is a remarkable process that goes beyond just washing away impurities. As the universal solvent, water can dissolve and transport a vast array of substances, from minerals and nutrients to toxins and pollutants. This unique property enables water to purify at both microscopic and planetary levels, sustaining life and promoting balance throughout ecosystems.

At a cellular level, water carries nutrients into cells and helps flush out waste, supporting vital biochemical reactions that keep the body functioning. In nature, water moves through landscapes, cleansing soils, replenishing rivers, and circulating essential minerals. Rainfall, streams, and groundwater carry away impurities, naturally recycling and renewing themselves in a process that nourishes life.

Just as water cleanses the environment, it mirrors processes within the human body. For instance, when we sweat, our bodies utilize water to release toxins and regulate temperature. Sweat is composed primarily of water, along with minerals and waste products, which are expelled from the body to maintain internal balance. This natural process of perspiration not only cools the body but also facilitates cleansing, similar to how water purifies the earth.

To maintain health and vitality, it is essential to drink healthy water every day. The solvent of the universe, water supports hydration, digestion, nutrient absorption, and detoxification. Clean water enhances cognitive function, energy levels, and overall well-being. By consuming pure and healthy water, we help our bodies operate efficiently, facilitating the processes necessary for life.

FUNDAMENTAL ROLE IN THE HUMAN BODY

An overview of how water plays a fundamental role in the human body, serving as a cornerstone for physiological functions and overall health. Water is an active medium that enables and optimizes the interconnected systems of the body. Without sufficient hydration, these processes can be impaired, leading to decreased performance, poor health, and long-term complications.

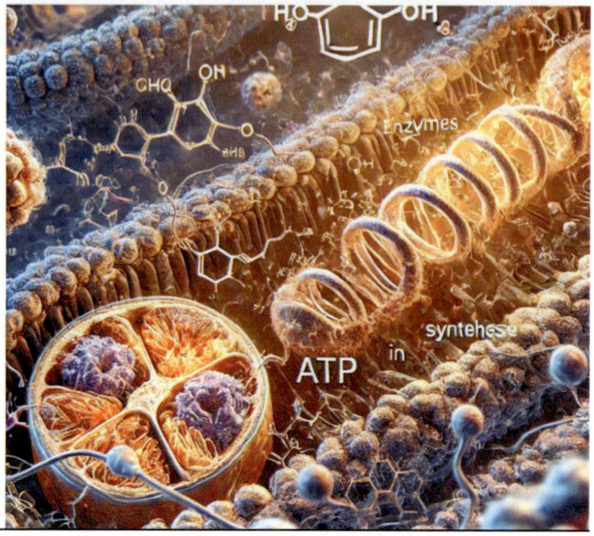

Hydrogen bonds in water can store information and influence cellular processes, highlighting the connection between water molecules and a human cell.

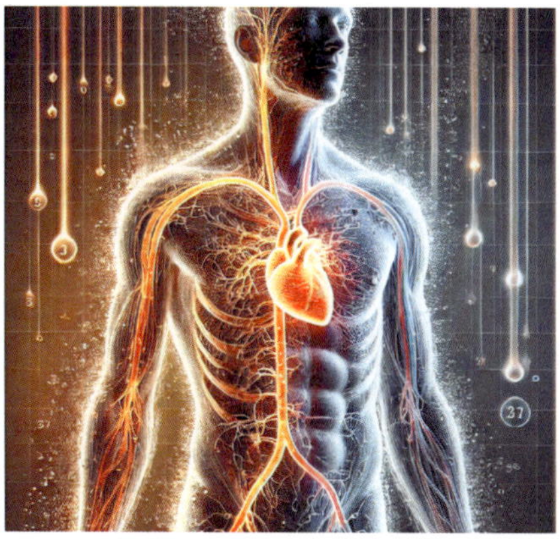 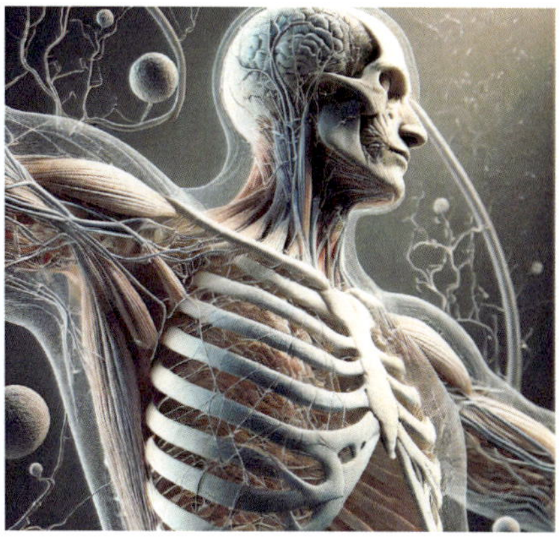 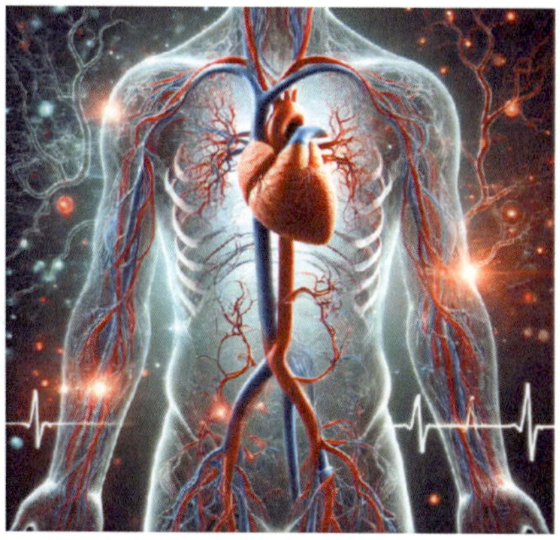

1. **Cellular Function and Hydration**
Water constitutes about 60-70% of the human body and is a critical component of cells, providing structure and maintaining cell shape. It acts as a medium for transporting nutrients, oxygen, and waste products to and from cells.
2.**Chemical Reactions:**
Many biochemical reactions, including those involved in metabolism, occur in an aqueous environment.
3. **Regulation of Body Temperature**
Water helps regulate body temperature through sweating and evaporation, maintaining homeostasis even in changing environmental conditions.
4. **Lubrication and CushioningJoints:**
It provides lubrication through synovial fluid, reducing friction between joints.
Tissues: Water cushions and protects vital organs and tissues, including the brain and spinal cord, by acting as a shock absorber.
5. **Digestive and Excretory Processes**
Water aids in breaking down food for nutrient absorption.
It plays a vital role in removing waste through urine and sweat, flushing toxins from the body.
6. **Blood Circulation**
Water is a key component of blood, making up plasma, which carries cells and nutrients throughout the body.
7. **Energy Production**
At the cellular level, water is essential for the production of energy (ATP) in mitochondria.
8. **Maintenance of pH Balance**
Water helps maintain the body's acid-base balance, ensuring that physiological processes function optimally.
9. **Communication and Signals**
In the nervous system, water enables the conduction of electrical signals, facilitating communication between cells.
10. **Immune Function**
Water supports the movement of immune cells and the transport of antibodies, aiding the body's defense mechanisms.
11. **Structural Support**
In connective tissues and the extracellular matrix, water contributes to the integrity and resilience of skin, cartilage, and other structures.

CHAPTER 3

WATER'S HIDDEN REALM: THE SCIENCE WITHIN

DR. MAE-WAN HO

THE LIVING RAINBOW: WATER'S QUANTUM DANCE

Dr. Mae-Wan Ho, renowned fo her book *"The Rainbow and the Worm: The Physics of Organisms"*, illuminated water's pivotal role in the quantum coherence of life. To Dr. Ho, water was far more than a chemical compound; it was the bridge between biology and physics, a quantum-coherent substance that linked cellular networks into a harmonious, unified system. Her work proposed that water enabled instantaneous communication within the body, orchestrating what she called a "quantum dance" among cells and tissues.

"Water is the most dynamic and connected element within living beings."

In *The Rainbow and the Worm*, Dr. Ho explored how the principles of quantum physics underlie the dynamic processes of living systems. She described organisms as liquid crystalline structures, with water playing a central role in maintaining this state of coherence. Water's unique properties allow it to form a continuous, interconnected network within tissues, transmitting information at the speed of light and enabling cells to function as a coordinated whole.

She saw life as a delicate interplay of light, energy, and water—a "living rainbow" of vibrant coherence. Through her research, she demonstrated that water's quantum properties are not just vital for communication within organisms but also for aligning the body's rhythms with the larger cosmos. In this way, water becomes a bridge, linking the microcosmic intricacies of cellular life with the vast, macrocosmic flow of universal energy.

Water is nature's delicate bridge, linking life's complexity through quantum coherence."

Water is not just a medium for chemical reactions; it is a conductor of life's symphony. Dr. Ho's work revealed that in living tissue, water behaves as a quantum-coherent matrix, transmitting energy and information with breathtaking precision. It enables biological systems to function in perfect harmony, aligning the organism's processes with the greater patterns of nature.

In The Rainbow and the Worm, Dr. Ho emphasized water's role in creating this coherence, describing it as a medium that bridges the complexity of life with the simplicity of physics. Water's quantum dance not only sustains life but reflects the interconnected nature of the universe, where every system—no matter how small—is part of a greater whole.

Water, as Dr. Ho articulated, is a living rainbow, a luminous thread that weaves life's fabric, connecting the smallest cellular movements to the cosmic rhythms of existence.

DR. ESTHER PATRICIA PEREZ DEL RIO

THE BODY AS A MAGNETIC, LIQUID-CRYSTAL SYSTEM

Dr. Esther Patricia Perez Del Rio, an expert in bioenergetics and biophysics, presents the human body as a complex, finely tuned system powered by magnetic fields and liquid-crystal water. According to her research, our bodies function like the most sophisticated computer in the world, where liquid crystalline water and magnetic forces work together to create a cohesive, self-regulating biological system.

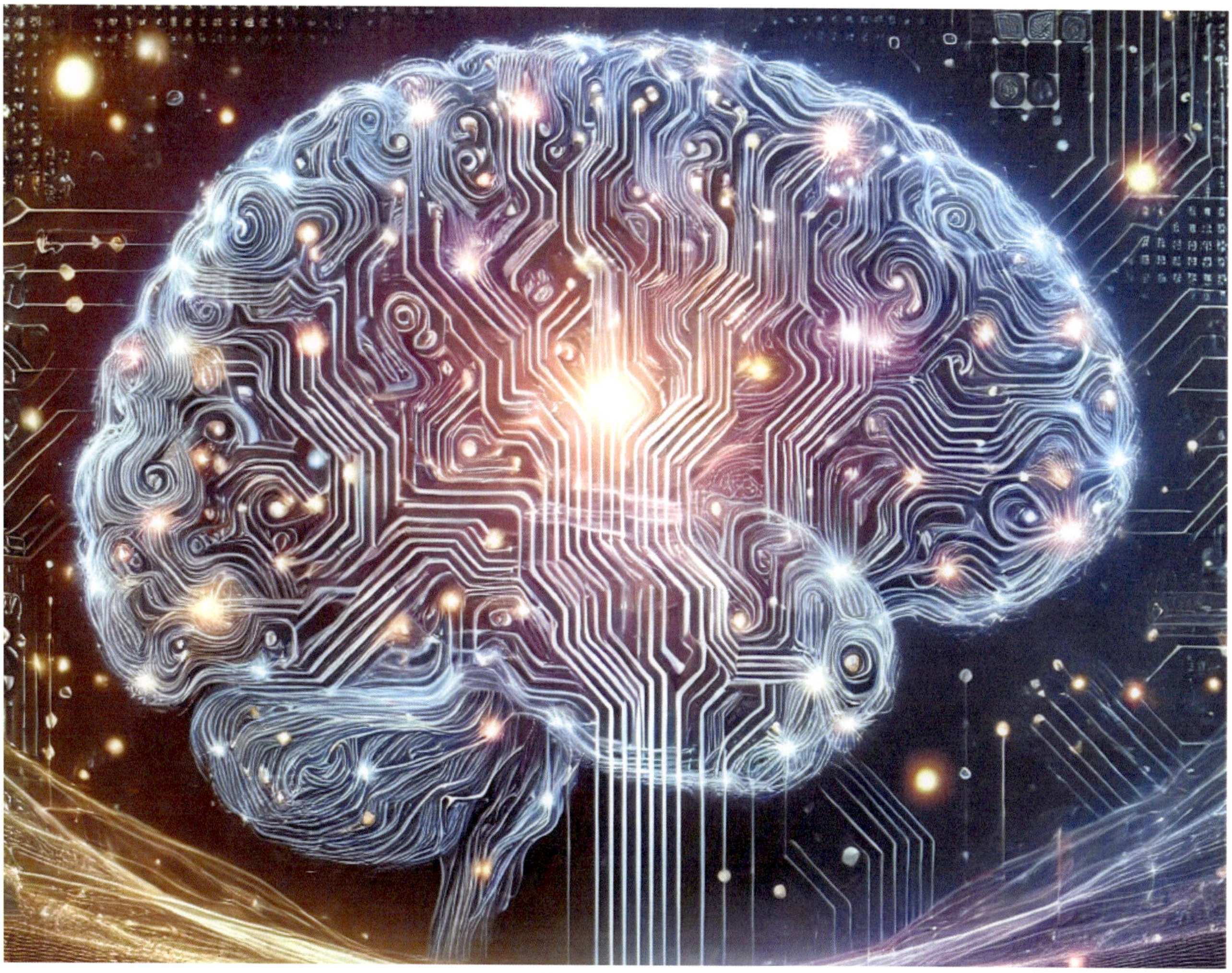

"Our bodies' magnetic and liquid crystal water create a biological system, making us the best computer in the world."

In Dr. Perez Del Rio's view, water is not just a solvent but an integral, dynamic part of our biological system, forming a structured, crystalline matrix that conducts and stores information. This liquid crystal water serves as a conduit for electrical signals, resonating with magnetic fields to maintain the body's internal harmony. Through this unique interplay of water and magnetism, our bodies not only function but also adapt and respond to the ever-changing environment around us.

The human brain, she suggests, is the ultimate internal hard drive. It doesn't merely process information but stores, orders, and records all sensory stimuli, experiences, and memories in a coherent, structured manner. The brain, working with the body's magnetic fields, uses this liquid-crystal water network to communicate instantaneously with every cell, coordinating our physiological responses with exceptional precision.

"Our brain's internal hard drive stores, records, and organizes all stimuli, creating an unparalleled biological computer capable of responding to the world around us."

This concept sheds light on the idea that the body's intelligence is both holistic and systemic. It reflects a sophisticated natural design where water, magnetism, and biological structure coalesce to form a responsive, dynamic living system. In Dr. Perez Del Rio's model, the body's natural magnetism, coupled with liquid-crystal water, enables a seamless flow of information, allowing us to experience and adapt to life in real-time with a complexity that no modern computer can replicate.

NIKOLA TESLA

THE LIVING AETHER: TESLA'S UNIVERSAL MEDIUM

Nikola Tesla, the visionary inventor and physicist, proposed that a universal "aether" pervades all space, acting as a subtle medium that connects and transfers energy and information throughout the cosmos. Tesla described this aether as a source of inspiration and knowledge–a universal field of energy that vibrates in waves, a concept resonating with the Greek philosopher Plato's notion of a boundless cosmic soul. This aether, Tesla suggested, could be the key to free, wireless energy if understood and harnessed correctly.

"My brain is only a receiver, in the Universe there is a core from which we obtain knowledge, strength, and inspiration. I have not penetrated into the secrets of this core, but I know that it exists."

From Tesla's perspective, this aether bridges matter and energy in a continuous flow, allowing for waves of electromagnetic energy to propagate and influence all forms of life. This aligns with recent studies on quantum fields, where the concept of "entanglement" suggests an invisible network connecting particles across vast distances.

Tesla believed that all life and consciousness connect to a central source of knowledge and energy. Water, with its unique ability to conduct energy and store information, can be seen as part of this universal network, bridging human consciousness with the greater cosmos.

Tesla's call for studying non-physical forces, such as energy and resonance, aligns with water's ability to capture and reflect vibrational information. This perspective encourages us to view water not just as a physical substance but as a bridge to the unseen forces Tesla spoke of.

"The day science begins to study non-physical phenomena, it will make more progress in one decade than in all the previous centuries of its existence."

DR. LUC MONTAGNIER

THE QUANTUM MYSTERY OF WATER AND BIOLOGICAL INFORMATION

Dr. Luc Montagnier, Nobel laureate and co-discoverer of HIV, expanded his scientific legacy by exploring water's potential to carry biological information. In his later research, Montagnier hypothesized that water could "remember" the electromagnetic signals of DNA, suggesting that water, as a medium, may encode and transfer information between biological systems in ways previously unrecognized by conventional science.

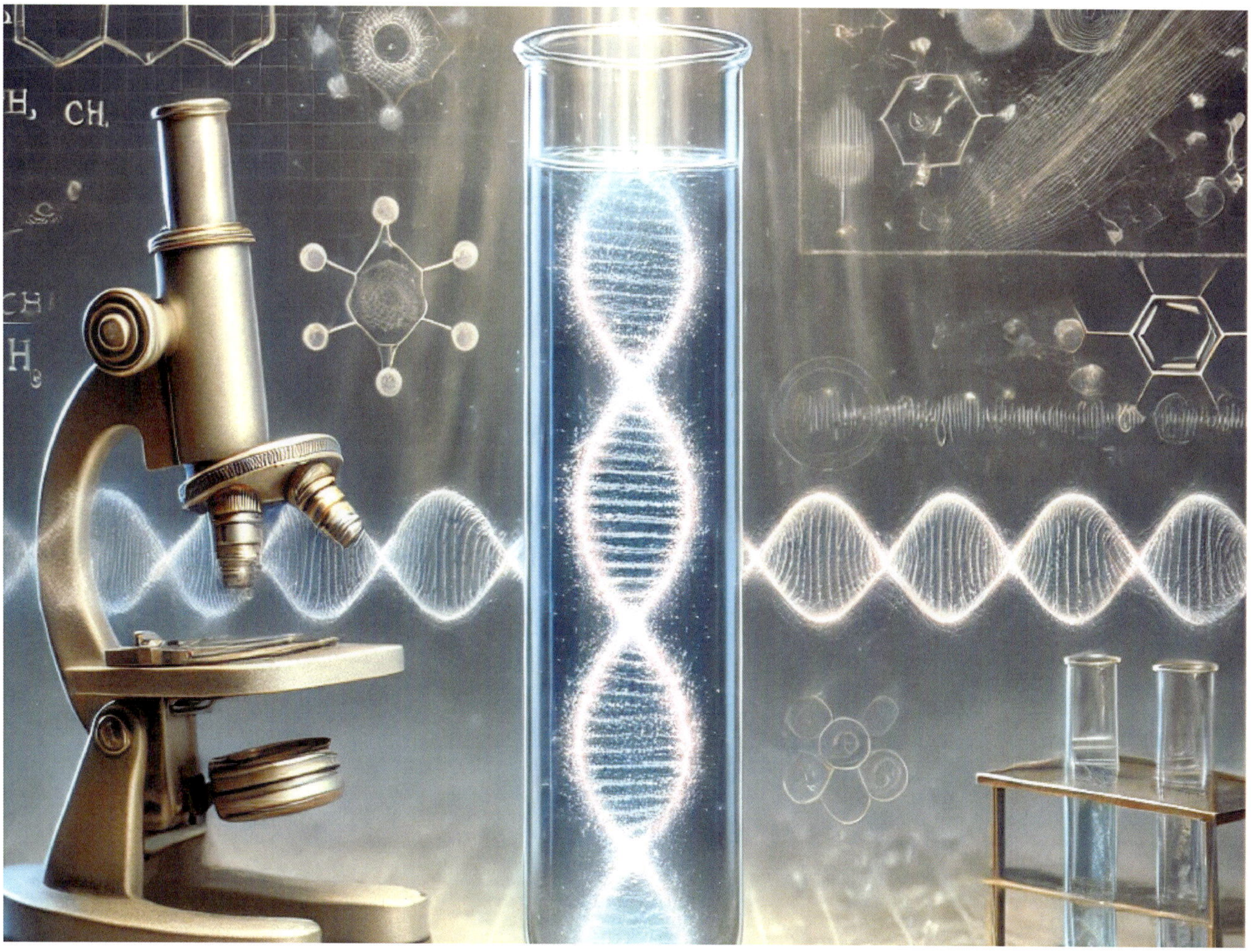

Water, as a medium, holds potential beyond simple hydration—it may encode and transfer information at a molecular level."

Montagnier's groundbreaking experiments revealed that highly diluted DNA solutions emitted specific electromagnetic signals, detectable even in the absence of actual DNA molecules. When these signals were transmitted to pure water samples, they appeared to reconstruct the original DNA structure remotely. This concept, though controversial, introduced the idea that water could act as a quantum bridge, storing and transmitting vibrational patterns that echo the original biological information.

Montagnier's work hints at a revolutionary understanding of biology, in which water's structured molecular network functions as a highly organized, dynamic system. This research bridges quantum theory and biology, suggesting that water may play an essential role in the cellular communication and memory that sustain life.

Through his explorations, Dr. Montagnier opened the door to new questions about water's properties, challenging us to reconsider the very foundations of biology and quantum physics. His contributions continue to inspire and provoke inquiry into the mysteries of water, energy, and the subtle forces that underlie biological existence.

"Water may act as a transmitter of life's fundamental information, bridging physical presence with quantum memory."

PROF. DR. BERND KRÖPLIN

THE BRIDGE BETWEEN THE IMMATERIAL AND MATERIAL

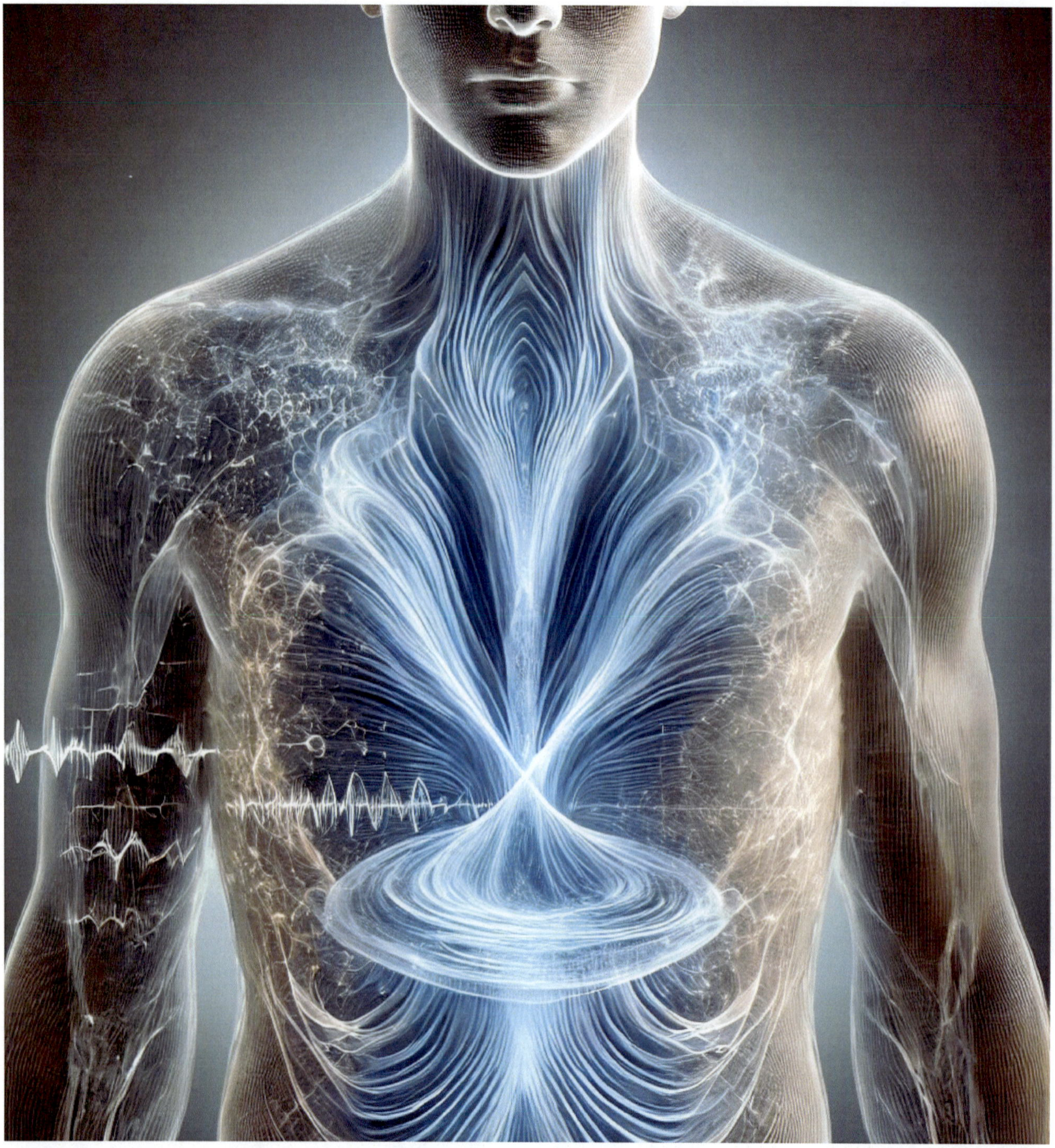

Kröplin, along with Regine C. Henschel and their team, has studied what he calls "the seven secrets of water." His research brings to light water's subtle responsiveness and its power to collect, store, and transport information, connecting us to a broader vibrational world. This work, though only beginning, offers a fascinating glimpse into water's capacity to serve as a connect between the tangible and intangible realms of existence, a silent recorder of the world around us.

Water holds more than just life—it holds memory, resonance, and information, fusing the immaterial with the material. Recent discoveries reveal that water "notes" and responds to external influences, capturing subtle imprints that shape its molecular structure. This capacity to remember and record its environment transforms water from a passive substance into a powerful medium of interaction between the unseen and seen worlds.

Through dark-field microscopy, scientists observe that after water evaporates, distinct patterns remain in the residue. These formations, often reproducible, reflect the information water was exposed to prior to evaporation, responding even to intangible factors like sound, electromagnetic fields, and chemical compounds. Water's structure shifts in reaction to these elements, displaying its unique ability to absorb and manifest vibrational influences within its molecules.

This is especially significant for us, as water constitutes approximately 70% of our bodies. The water within us collects, holds, and transports information, influencing cellular health and our overall resonance. Whether exposed to music, light, or various frequencies, water adapts, attuning itself to environmental conditions and recording the "frequency footprint" of our experiences.

As the carrier of resonance, water transfers this information through our cells, making it a critical player in the delicate balance of our biological systems. These imprints, from microscopic levels to macro scales, show how water is continually receiving, transforming, and releasing energy, engaging in the cosmic cycle of giving and regiving. In this way, water reminds us that, though it may change forms, it remains connected to all life—a journey of oneness in many forms.

"Water notes and remembers the external influences upon it. Each molecule acts as a bridge between the immaterial and material worlds, storing these subtle phenomena within its structure and allowing us to glimpse the unseen forces shaping our reality."

LIQUID TAPE RECORDER

PROF. BENVENISTE, DR. LUDWIG AND PROF. SCHWEITZER

The work of Prof. Benveniste, Dr. Ludwig and Prof. Schweitzer has demonstrated how water acts as a liquid tape recorder. Water stores and transmits electromagnetic vibrations. This emphasizes the fact that we are our thoughts spoken in words.

Water as a liquid tape recorder is not just a metaphor–it is a profound truth reflecting the interconnectedness of all life. Its ability to store, transmit, and amplify vibrations underscores its role as the medium of memory, life, and the cosmos. This idea bridges ancient wisdom, modern science, and spiritual insight, offering a deeper understanding of water's sacred and practical significance

Molecular Structure: A Dynamic Matrix

- Water's polarity and ability to form hydrogen bonds create a dynamic network that can shift and reorganize.
- This network allows water to act like a liquid crystal, adapting to and recording information from its environment through changes in its structure.
- The clusters of water molecules can temporarily organize themselves in patterns, influenced by external stimuli such as sound, light, or electromagnetic fields.

Storing Electromagnetic Vibrations

- Water molecules respond to electromagnetic frequencies by aligning and realigning their dipole moments (positive and negative charges).
- This alignment can leave an imprint or "memory" of the frequency, even after the original source is removed, much like how magnetic tape stores sound.
- In biological systems, water is constantly interacting with the body's electromagnetic fields, storing and transmitting cellular information.

Transmitting Vibrations

- Water's ability to transmit vibrations lies in its fluidity and conductivity.
- Vibrations propagate through water as pressure waves, enabling it to act as a medium for transferring energy and information over distances.
- In the human body, water transmits vibrational signals between cells, facilitating communication and maintaining coherence in biological processes.

Applications in Nature and Technology

- Biological Memory: Water in the body plays a crucial role in storing information at the cellular level, impacting DNA expression and health.
- Energy Medicine: Water is used in therapies like homeopathy, where it is believed to retain the vibrational essence of substances.
- Quantum Information Transfer: Some scientists theorize that water's structure could enable quantum coherence, allowing it to transmit information at a fundamental level.

Water as a Holographic Recorder

- Holographic Principle: Like a hologram, water can store a vast amount of information in a small space, encoding the essence of the larger system.
- Feedback Loop: Water records environmental vibrations and gives feedback through resonance, creating a loop of continuous interaction.
- Memory of the Universe: As water cycles through the planet, it absorbs the vibrations of life, earth, and cosmos, becoming a living archive of the world's history.

DR. MASARU EMOTO

THE REFLECTIVE CONSCIOUSNESS OF WATER

Dr. Masaru Emoto's research took this sensitivity further, showing how water could respond to human intention. Emoto exposed water samples to positive and negative words, then froze and examined them under a microscope. *"Water reflects our consciousness,"* he would say. The crystals formed by water exposed to loving thoughts were beautifully symmetrical, while those exposed to negative thoughts became chaotic. His work suggested that water not only reflects physical properties but also responds to emotions and intentions, acting as a mirror to the human spirit.

"Water reflects our consciousness."

Dr. Emoto's groundbreaking work in crystallography showed that water responds to human thought and intention. He discovered that positive words and feelings, such as "love" or "gratitude," produced beautiful, symmetrical crystals in frozen water, while negative thoughts created chaotic structures. Emoto saw water as a mirror to the human soul, reflecting our emotions and intentions.

As Emoto relates, "If you shine a light on those around you with the words you use, you won't ever have to walk in the dark."

Emoto's findings suggested that water is sensitive to human consciousness, a substance that not only sustains life but also reflects the spirit's inner state.

DR. KONSTANTIN KOROTKOV

WATER AS AN ENERGETIC IMPRINT AND RECORDER

Dr. Konstantin Korotkov, a Russian scientist and professor, is renowned for his pioneering work in electrophotonic imaging and research on water's energetic properties. He has been instrumental in developing the Gas Discharge Visualization (GDV) technique, a method that captures the bioenergetic field of living organisms and even water, showing how water can hold energetic imprints from various influences. Korotkov's work delves into how water not only sustains life but also serves as a record-keeper, holding energetic and informational imprints from its surroundings.

 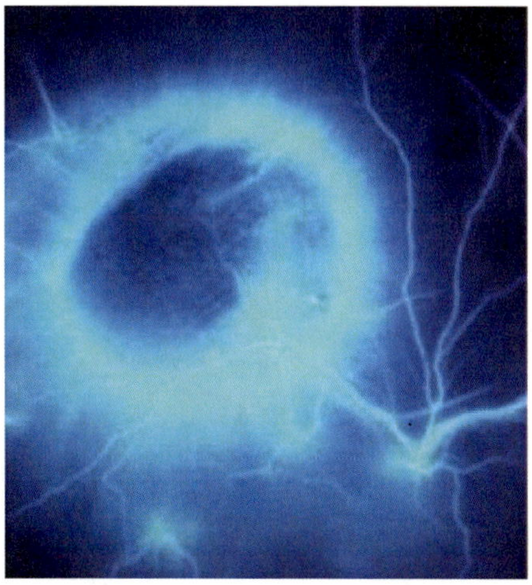 

Dr. Korotkov's work with GDV imaging, bio-grams, and his studies on water's imprint abilities contribute to a growing body of research on water's role as both a life-supporting substance and a record of energetic and informational influences. His findings reinforce the concept that water acts as a bridge between physical and energetic realms, able to absorb, store, and transmit information from the world around it.

Korotkov's research shows that water's molecular structure allows it to record energy from its environment, much like a biological memory. His studies with GDV imaging reveal that water exposed to different influences–such as thoughts, emotions, or electromagnetic fields–exhibits distinct energetic patterns, which he believes are visible imprints of these exposures.

"With Gas Discharge Visualization, we can observe the life force or biofield that water possesses, revealing its dynamic interactions with energy and information."

GDV, also called Electrophotonic Imaging, captures the biofield or "aura" around water samples, showing how water responds to different stimuli. Korotkov found that structured or "energized" water emits a more vibrant and organized biofield, which he suggests reflects its higher energetic quality and coherence.

"Water is the ultimate observer—it reflects and remembers everything it encounters, holding the imprint of the physical and energetic worlds."

Korotkov believes water is an "observer" of the universe, capable of recording and retaining the essence of the environments and energies it encounters. This perspective aligns with the idea of water as a sensitive recorder, where even subtle vibrations or thoughts leave a lasting mark on its structure.

For Korotkov, water's ability to carry energetic imprints makes it a connector between the physical metaphysical realm. He has demonstrated that water's structure can reflect energetic patterns that influence living organisms, making it a vital link between energy and biology.

Korotkov's work with GDV imaging shows that each water sample has a unique energetic "fingerprint," impacted by various factors in its environment. This fingerprint is thought to be a result of water's interaction with its surroundings, preserving its energetic history.

"Water is an extraordinary substance; it holds memory and has the ability to absorb, store, and transmit information."

CHAPTER 4

CRYSTALLINE LATTICE, TABLE OF ELEMENTS, COLOR OF LIFE & WELLNESS

CRYSTALLINE LATTICE OF WATER

A SACRED GEOMETRY OF RESONANCE

"All life on Earth is composed mostly of carbon-12 and water. This is the case because carbon-12 bonds resonate with more simple elements than any other element in the universe. It is for this very reason that carbon-12 is the international standard for atomic weight and all other elements are measured against it. With 6 protons + 6 neutrons in its nucleus and 6 electrons orbiting in two shells, carbon-12 exhibits the lowest possible energy of all the elements and is said to be 'unbound,' thereby creating the most stable atomic geometry possible. When mixed with water, carbon-12 creates endless chains of sticky amino acids capable of crystallizing into life.

"This idea of life as a crystallization process is a good one because just as minerals align under pressure into lattices, coils of amino acids fold under pressure into three-dimensional protein structures, aligning into the familiar helical lattice of DNA. It is the geometric pressure of hydrogen atoms in water that helps create the lattice and give DNA its twist." ~ Richard Merrick, Interference Theory

Water's three-dimensional tetrahedral lattice is more than just a structure; it is a finely tuned matrix for carrying encoded resonance and harmonic ratios. In two-dimensional representations, as illustrated (right) by Martin Chaplin Ph.D., water clusters form intricate arrangements–280-molecule clusters that align oxygen and hydrogen atoms in patterns that echo sacred geometries like Platonic solids and the Flower of Life. These geometries reflect water's ability to act as a conductor for vibrational information.

Water's three-dimensional tetrahedral lattice is more than just a structure; it is a finely tuned matrix for carrying encoded resonance and harmonic ratios. In two-dimensional representations, as illustrated by Martin Chaplin Ph.D., water clusters form intricate arrangements–280-molecule clusters that align oxygen and hydrogen atoms in patterns that echo sacred geometries like Platonic solids and the Flower of Life. These geometries reflect water's ability to act as a conductor for vibrational information.

Gene Webb expands on this by suggesting that water's lattice structure may parallel the mechanisms of gravitational lensing in dark matter, which bends light across the cosmos. Water, in this view, becomes a vessel for resonance ratios–holding and transmitting the fundamental frequencies of the universe.

Life on Earth, composed mostly of carbon-12 and water, mirrors these harmonic principles. Carbon-12, with its atomic geometry, offers stability and resonance with a vast array of elements, making it the perfect counterpart to water's fluid structure. Together, water and carbon-12 create an environment in which amino acids crystallize under pressure, forming helical structures like DNA. Just as minerals align into lattices, coils of amino acids fold into three-dimensional protein structures, giving rise to the complex architecture of life. Merrick notes,
"It is the geometric pressure of hydrogen atoms in water that helps create the lattice and give DNA its twist."

This idea aligns beautifully with concepts like E-8 theory, (Theory of Everything) which explores how intricate symmetry underlies the universe. Just as E-8 represents a highly symmetrical and complex structure in mathematics and physics, water's lattice may operate as a living example of this symmetry, interfacing with the quantum field and echoing the resonance of cosmic geometry. it also colorates to Russell's crystlline chart of light.

Water's lattice, in essence, bridges the physical with the energetic, shaping and sustaining life through the very principles that govern the fabric of the universe.

"...We now know that strongly interacting electrons in graphene (carbon) behave like a liquid. Materials conduct heat in two ways: through vibrations in the atomic structure or lattice; and carried by the electrons themselves."
~ Peter Allen, Harvard SEAS

THE JOURNEY OF A DROP

A BRIDGE BETWEEN DISCIPLINES AND PERSPECTIVES, SHOWING THEIR UNITY IN THE GREATER TAPESTRY OF EXISTENCE

A single drop of water is far more than a simple molecule; it is a microcosm of the universe, reflecting the principles of life and creation. Like the heart, a drop of water embodies polarities–positive and negative charges–that generate an oscillating flow of energy. This interplay acts as an electrical pump, mirroring the rhythmic pulse of the human heart. Each beat propels life forward, just as the charged dynamics within a drop of water circulate energy and vibration, sustaining the flow of life.

"In the grand design, no drop is lost, for all are threads in the infinite fabric of existence." ~ Inspired by Nikola Tesla

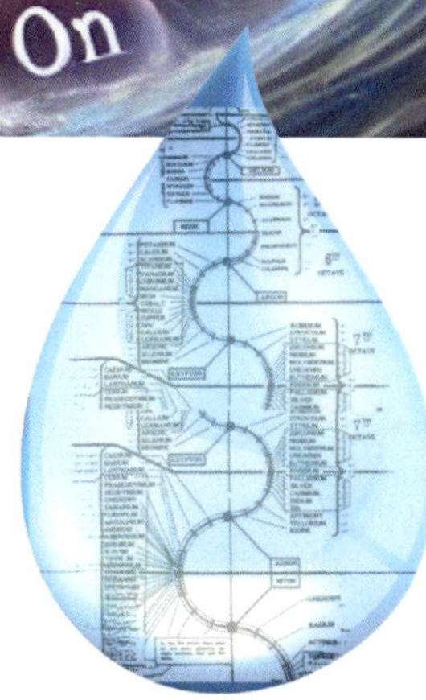

Water resonates across the full spectrum of octaves, from the deepest bass to the highest treble, making it the universal conductor of the symphony of creation. Each octave is a layer of vibration, an energetic rhythm that encodes the water with the harmonic resonance of existence itself. Within every drop, this universal heartbeat can be felt, connecting the microcosm to the macrocosm –the part to the whole.

"Creation is but a ripple in the stillness of the infinite ocean of God's light."
Walter Russell

The structure of the water molecule itself reflects universal harmony. As Babbitt's atom model suggests, the relationships between atomic elements create a cascading dynamic, much like the heartbeat of the universe. When combined with Walter Russell's cascading periodic table, represented as nine octaves of creative energy, we see how water exists as a harmonic medium–perfectly tuned to the universal scale. Each octave in this table represents a vibrational layer of creation, flowing like a wave, uniting light, matter, and energy. Water, the element of life, becomes a vessel of these octaves, its very structure reflecting the geometry of the cosmos.

This mirrors Viktor Schauberger's concept of the cosmic egg, where the universe itself can be seen as a dynamic, spiraling vortex. Schauberger observed that water moves naturally in a spiral motion, spinning its web via the golden ratio, Phi, generates life's energy. Within this spin lies the key to creation itself–a self-regenerating force that mirrors the toroidal motion of galaxies, magnetic fields, and even the human heart. The cosmic egg, with its universal spin, reflects the generative principle of all creation, from the largest galaxies to the smallest drops of water.

"We are but whirlpools in the cosmic river of life. Each twist, turn, and ripple is a reflection of the universal flow, as above, so below, so within." ~ Inspired by Viktor Schauberger

On a planetary scale, the oceans echo this same principle. The great ocean conveyor belt–Earth's circulatory system–propels water and nutrients across vast distances, mirroring the heart's function within the human body. Just as the heart's electrical charge governs the flow of blood, the oceans are driven by the magnetic field of the Earth, creating currents that circulate energy, sustain ecosystems, and regulate planetary balance. This magnetic interplay is a reflection of the scaling law: what is true in the smallest drop is equally true on the largest scale.

"Don't you realize that the sea is the home of water? All water is off on a journey unless it's in the sea, and it's homesick, and bound to make its way home someday." ~ Albert Szent-Györgyi

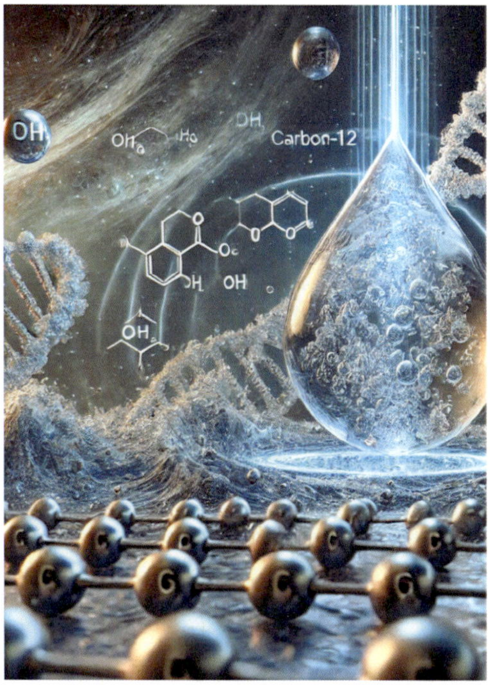

From the primordial soup—the birthplace of life—this interconnected dance began. Charged particles within water aligned with magnetic fields, forming the first seeds of life. Water, as the cradle of creation, harnessed the vibrations of the cosmos, bridging matter and energy. The cascading periodic order within water, informed by its nine octaves, mirrors the creative principle of life itself. The same forces that gave rise to the first cells continue to sustain life today. In every drop of water, the memory of this origin is imprinted, a living archive of the universe's creative force.

"A molecule of water is a galaxy in miniature, a cosmos held in perfect harmony."- Inspired by Gerald Pollack

And when a drop of water falls into the ocean, it is no more connected than it was in its station as a single drop. Creation has no escape from the whole. Our connection to this relational reality is the key to our ability to relate fully with the whole. We could shed a tear of joy, and the full knowledge of the universe would be embedded in its structure and record, a reflection of the infinite in the finite.

"We are waves on the cosmic ocean, each unique, yet inseparable from the infinite whole."- Inspired by David Bohm

We are but drops in the ocean, yet inseparable from the whole. Just as a single drop contains the same vibrational harmony as the vast seas, we carry within us the same operating principles as the cosmos.

The rhythmic pulse of water—whether in a drop, an ocean current, or a cosmic wave—unites all life into a singular, magnificent flow.

Thus, a drop of water is not merely a substance but a sacred vessel—a heartbeat of the universe itself. It reflects the unity of existence, scaling from the smallest fractal to the largest system. Each ripple, each wave, and each heartbeat reminds us that we are a part of something far greater, a living symphony where every drop contributes to the song of creation.

"Water flows not only on the Earth but within us, connecting our essence to the heartbeat of the cosmos."- Unknown

RANDY HATTON

WATER AS A PROGRAMMABLE LIQUID CRYSTAL

Randy Hatton, a leading expert in water science, explores the remarkable crystalline nature of water at the molecular level, uncovering its dynamic and adaptive qualities. Hatton posits that each water molecule contains up to 440,000 informational panels, and there are over 1.5 sextillion molecules in a single drop—a number followed by twenty-one zeros. According to his pioneering research, the most healing and beneficial water to consume and bathe in is what he calls Full-Spectrum Water.

By creating a free-flowing centripetal interwinding vortex—or toroid—in cold water, we can efficiently implode a variety of gasses. When combined with sound frequencies, full-spectrum light, magnetism, paramagnetism, specific scalar waves and a balance of full spectrum ocean minerals, this process transforms water into a medicinal form, "Full-Spectrum Water."

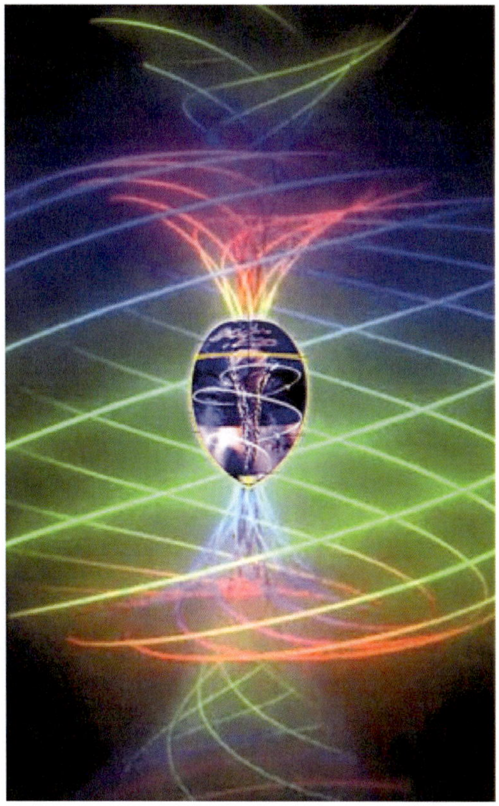

Beyond the influence of light, Hatton highlights the profound significance of a Hyperbolic egg shape, which supports creation of the ultimate toroidal motion in water. This dynamic reflects the forces of both centripetal implosion and centrifugal expansion—two opposing yet complementary movements happening simultaneously. These forces mirror the balance found in nature, aligning with the universal law of polarity: the yin and yang, feminine and masculine, hot and cold, and even the hemispheres of the brain—right and left.

This dynamic interplay, Hatton suggests, is critical for restoring homeostasis, the natural energy balance of our planet and within ourselves. As Walter Russell described through the "sex principle" in creation, it is the dance of opposing forces that allows life to emerge and flourish. Without harmony between these polarities, creation and evolution cannot occur. In an age where imbalance dominates, Hatton emphasizes the urgent need for humanity to return to the equilibrium of these natural forces.

Hatton's purpose is to help humanity reconnect with the spirit of water, treating it as a sacred companion–with love, respect, and honor. By mimicking nature's intricate geometries and patterns, we can deepen our relationship with water and awaken to its extraordinary potential. He suggests that water's structure is "programmable" by various environmental factors, including light and color. The interaction of light with water reveals an extraordinary adaptability, as different colors influence water's energy, structure, and even its chemical behavior. Hatton explains:

"It has become very clear that different colors of light on water will affect it in different ways."

This interaction leads to subtle changes in water's crystalline patterns and molecular behavior, allowing it to dynamically respond to its environment.

Oxygen's molecular structure further amplifies this adaptability, resonating with various frequencies across the spectrum–relating to octaves in sound and energy. Oxygen's ability to bind and hold energy enhances water's capacity to absorb, retain, and transmit energetic patterns, making it a universal medium for resonance. This property is especially critical in biological systems, where water's unique structure–particularly in its fourth phase, or Exclusion Zone (EZ) water–facilitates cellular communication, energy transfer, and coherence across life processes.

Water, infused with informational fields through pulsating light, becomes even more potent. Different colors and elements produce unique energetic imprints within water:

- Red hydrogen
- Blue oxygen
- Green nitrogen
- Magenta potassium
- Scarlet magnesium

Each combination amplifies water's healing and life-sustaining properties, transforming it into a vibrant medium for energy and balance.

Quantum jazz is played out by the whole organism, in every nerve and sinew, every muscle, every single cell, molecule, atom, and elementary particle, a light and sound display that spans 70 octaves in all the colors of the rainbow.
- Dr. Mae-Wan Ho

Chemical interactions also play a critical role in shaping water's vibrational character. Water's molecules create unique vibrational signatures that can support cellular processes, ecological harmony, and even states of consciousness. The oxygen molecule within H_2O interacts dynamically with light, sound, and energy, making water a programmable medium that integrates light, chemistry, and structure into a single, harmonious system capable of sustaining life on both physical and energetic levels.

Water is, indeed, the source and mother of all life. It flows through every aspect of existence, connecting the elements of Earth, supporting ecosystems, and serving as a bridge between science and spirit. Through Hatton's perspective, water becomes a mirror to the universe—responsive, intelligent, and alive.

By treating water with reverence and understanding, we unlock its full potential to heal, transform, and nurture the living world. In doing so, we align with the deeper principles of natural law: balancing forces, honoring the dance of creation, and embracing water as both a guide and a gift on our journey toward harmony and homeostasis.

Water's molecular makeup an entire "spectrum" or "octave" of resonances, bridging the gaps between chemistry, biology, and physics, allowing it to support life on both physical and energetic levels. It's an inspiring way to think about water's place in the universe! This insight was suggested by Don Estes.

WALTER RUSSELL

THE COSMIC BLUEPRINT OF LIFE

Walter Russell, the visionary artist and scientist, presented a revolutionary approach to understanding the cosmos through his wave-based periodic table. For Russell, all elements are part of one fundamental Cosmic Mind, expressed across varying densities of light, water, and energy within the continuous wave of creation. His table visualizes elements as interconnected manifestations of a single element, existing in harmony through the cosmic wave's rhythmic flow.

Russell's periodic table organizes the elements not as isolated entities but as dynamic stages within the unified wave, where every element, from the lightest gas to the densest solid, is but a different density of the same essence. This structure reveals the universe as a seamless, undivided field of energy, or Mind, where light, water, and matter are interconnected facets of one element expressing itself across the cosmic scale.

"To know the mechanics of the wave is to know the entire secret of the universe."

In Russell's philosophy, the Cosmic Mind is the ALL, the essence that flows through every atom, wave, and particle. The periodic wave is a reflection of this singular, boundless Mind, where every element resonates with the same creative impulse, echoing the principle *"as above, so below, so within."* This cosmic unity illustrates that all elements, regardless of form, are part of an unbroken continuum–differentiated only by their density within the cosmic wave.

Russell's holistic model in his book "The Universal One" aligns the microcosm and macrocosm, where light, water, and every particle serve as reflections of the whole. The wave reveals that each density of matter and energy is a lane within the Cosmic Mind, embodying a higher order that unites physics, spirituality, and consciousness. By understanding this wave-based periodic table, we can begin to see ourselves not as separate from the cosmos, but as vessels of the same rhythm, life, and creative energy that sustains the stars.

"The universe is a symphony of waves within the Cosmic Mind, where light, water, and all matter dance as one unified element."- Walter Russell

FROM CHAOS TO CREATION

THE BIRTH OF THE UNIVERSE

We are HUE-MAN, woven from the colors of the rainbow, each of us a unique shade on the spectrum of light. Born into the ocean of existence, we are like vibrant eddies, swirling in the vast sea, each ripple a distinct expression of life's infinite possibilities. The hues we carry are not only colors but frequencies, vibrations that shape who we are and how we resonate within the universe.

Every color, every hue, is a note in the cosmic symphony, and each of us brings a different tone to the harmony of existence. We are both the ocean and the wave, both the light and the color it reveals, reflecting and refracting through the prism of life. United in diversity, we flow together in the endless dance of being, each hue a unique stroke in the masterpiece of creation.

As Walter Russell beautifully stated: *"The wave is the cosmic stage, the light is the actor, and the play is the drama of creation."*

From the crystalline river of life, flowing from the heavens, comes the fundamental pulse of existence–vibration. This river, woven from light and energy, vibrates like the strings of a harp, creating harmonies that echo through the universe. Just as the river of life flows, so too does the vibrational pulse that shapes and sustains all creation.

In Revelation, we are told of the 12 branches and 12 fruits, symbolizing the divine harvest and the eternal connection between all living things. These branches extend from the tree of life, each one a unique expression of the divine flow. The fruits they bear represent the gifts of creation, each fruit ripening in its own time, yet all nourished by the same source.

"And on either side of the river was there the tree of life, which bare twelve manner of fruits, and yielded her fruit every month: and the leaves of the tree were for the healing of the nations." ~ Revelation 22:2

This vivid image connects the crystalline river to the very heart of existence, where the vibrational energy of creation sustains not only the cosmos but the very life within it. The river of life, flowing from the heavens, is the source of these vibrations–creating the harmonic symphony of existence, nourishing and healing all that it touches.

The 12 branches stretch across the universe like the arms of the divine, each one resonating at its own frequency, each fruit a unique note in the cosmic symphony. And just as water carries light and energy, so too do the waves and vibrations of this river carry the drama of creation, with each wave representing a new chapter, each ripple an unfolding of life's mysteries.

"That which man calls matter, or substance, has no existence whatsoever. So-called matter is but waves of the motion of light, electrically divided into opposed pairs, then electrically conditioned and patterned into what we call various substances of matter. Briefly put, matter is but the motion of light, and motion is not substance. It only appears to be. Take motion away and there would not be even the appearance of substance." ~ Walter Russell, The Universal One

THE LAW OF TRIANGLES

TETRAHEDRON GEOMETRY, LIGHT, WATER, AND ATOMS

At the core of all creation lies the Law of Triangles, the sacred geometry of life's angles. This principle governs the foundational structures that shape existence. Born from the sinusoidal and cosine waves, the interaction of these fundamental oscillations gives rise to intricate patterns forming the universe's very fabric. These waves evolve into geometrical structures, shaping all that exists–from the smallest subatomic particles to the vastest cosmic entities.

"Each human being is a prism, refracting the light of the cosmos into the unique colors of their soul." ~ Anonymous

At the heart of this geometry is the tetrahedron, a three-dimensional triangle and the building block from which all Platonic solids are formed. These solids configure life itself, embodying the cosmic dance of balance, symmetry, and energy. As light, water, and atoms bond through the power of three, they create the triadic foundation of existence, each element resonating in harmony to form the fractal patterns that define life's structure. This triadic nature, woven into the cosmos, echoes across every atom, cell, and star, reminding us of the interconnectedness of all things.

As cosmic waves interact with the tetrahedral geometry, light passes through the prism of creation, refracting into a spectrum of color and frequency. These frequencies vibrate within all things, illuminating existence itself and revealing the infinite fractal patterns emerging from this sacred dance. The "treeing" effect–energy branching and spiraling in intricate, self-replicating forms–reflects this fractal geometry, echoing the growth of a tree, with each branch maintaining the same geometrical harmony at different scales.

This evolution of matter, from aether to water, reveals light, water, and atoms bonding to create life's framework. The energetic pulse of this process reverberates throughout the cosmos, guiding and shaping creation and unveiling the profound interconnection between all things.

"The cosmos is within us. We are made of star-stuff. We are a way for the universe to know itself."– Carl Sagan

In this cosmic blueprint, the tetrahedron stands as the building block of the Platonic solids from which all life is configured. As Johannes Kepler suggested, "The universe is a living image of God," reflecting an order that transcends even our perception–a universe woven from light, geometry, and energy into a coherent whole, echoing the divine architecture of existence.

DON ESTES

VIBRATIONAL MEDICINE AND WATER AS A CONDUIT FOR CONSCIOUSNESS

Don Estes is a pioneer in the field of vibrational science, consciousness, and resonance. Known for his work with bio-resonance technology and concepts around vibrational medicine, Estes explores the relationship between sound, vibration, and water's role in human consciousness and health. He has developed technologies that use sound frequencies to influence water within the human body, promoting physical and mental wellness. Estes believes that water, when activated through vibrational energy, can serve as a powerful medium for healing and consciousness expansion.

Don Estes' work with vibrational science emphasizes water's unique role as a resonant medium that bridges physical health and consciousness. His belief in water's ability to store and transmit energy aligns with the principles of vibrational medicine, suggesting that sound, light, and frequency can transform water within the body, promoting holistic well-being and expanded states of consciousness.

Estes sees water as a "living antenna" that can carry vibrational information, resonating with different frequencies to promote healing. He believes that water's natural structure makes it especially sensitive to vibrations, allowing it to act as a carrier of energy and information.

"Every cell in the body is like a musical instrument, tuned to a specific frequency, and water is the medium through which these frequencies communicate."

Estes compares the body's cells to musical instruments, each resonating at specific frequencies. Water, he suggests, is the conductor of this biological orchestra, facilitating communication among cells and helping them maintain harmony.

"Vibrational medicine is about aligning with nature's inherent frequencies. By tuning into these frequencies, we can create coherence in both body and mind."

Estes' approach to vibrational medicine emphasizes the idea that natural frequencies, when introduced to the body, can bring cells and systems back into alignment. He sees water as a fundamental element in this process, a medium that can absorb, carry, and amplify healing vibrations.

"When sound and light frequencies interact with water, they create a resonant field that can alter consciousness and promote wellbeing."

Estes believes that sound and light frequencies have profound effects when transmitted through water. This interaction creates a resonant field that can enhance states of awareness, induce relaxation, and stimulate healing on both a physical and psychological level.

For Estes, water serves as a "gateway" to higher consciousness. Through vibrational tuning, water can store energy and facilitate experiences of expanded awareness, bridging the physical body with the mind's higher states.

GARY ROBERT BUCHANAN

THE BODY AS A LIVING HOLOGRAPH OF LIGHT AND SOUND

Gary Robert Buchanan was a pioneering researcher in vibrational science using Sonatherapy, a therapeutic approach integrating light, color, sound, water, and subtle energies to promote holistic healing. Buchanan's work explores the human body as a "living holographic entity" composed of light (photons) and sound (phonons), suggesting that this dynamic interplay supports physical and energetic well-being. Through Sonatherapy, he introduces innovative concepts on how vibrations and resonance within the body's fields facilitate healing, emphasizing the body's role as a conduit of universal frequencies.

Gary Robert Buchanan presented a revolutionary view of the human body as a "living holographic entity"–an intricate tapestry of light and sound in motion. Buchanan proposes that all phenomena, including the human body, exist as patterned vibrational states, where sound (phonons) and light (photons) harmonize in a constant, dynamic flow.

This holographic nature of the body is more than an image; it is a record of the light from which we are composed, each cell and vibration a memory of the spectrum that creates us. Sound and light interact continuously, forming waves of resonance that shape our energetic field. Buchanan described "propagated phonon chains"–the sounds, songs, and vibrations within our environment–as pathways that deepen our ability to absorb and emit photons, thus forming visible expressions of energy. The essence of our being emerges from this constant, harmonious interplay between light and sound, as photons flow in response to the vibrational pathways created by phonons.

Through Sonatherapy, Buchanan introduced the idea that light and sound are more than physical structures–they are conduits that nourish, sustain, and heal. This approach bridges the physical with the energetic, portraying the body as a "holograph of light and sound" woven from the frequencies of the universe. Each wave, whether a note, color, or spoken word, resonates within this holographic field, enriching our experience, creating our presence, and contributing to the symphony of life.

Dr. Samuel Hahnemann, a pioneering German physician and pharmacist, introduced the world to homeopathy, a natural medical system grounded in two foundational principles: the law of similars and potentization. According to Hahnemann's law of similars, substances that cause symptoms in a healthy person can, when specially prepared, heal similar symptoms in an unwell person.

"Water remembers; it holds the essence of all it encounters, echoing the vibrational patterns of life."

SAMUEL HAHNEMANN

FOUNDER OF HOMEOPATHY AND THE PRINCIPLE OF POTENTIZATION

Hahnemann developed the concept of potentization, a process of extreme dilution combined with vigorous shaking, or succussion, which he believed enhanced the remedy's healing efficacy. This process, Hahnemann argued, "unlocks dormant, hidden medicinal powers" within the substance, enabling it to act at a profound, energetic level. Contrary to the conventional belief that dilution weakens a substance, he proposed that potentization actually amplifies the healing potential of the original material.

"Homeopathy is more than mere dilution; through potentization, we unlock the subtle energies within, revealing nature's innate power to heal."

Hahnemann developed the concept of potentization, a process of extreme dilution combined with vigorous shaking, or succussion, which he believed enhanced the remedy's healing efficacy. This process, Hahnemann argued, "unlocks dormant, hidden medicinal powers" within the substance, enabling it to act at a profound, energetic level. Contrary to the conventional belief that dilution weakens a substance, he proposed that potentization actually amplifies the healing potential of the original material.

"Homeopathy is more than mere dilution; through potentization, we unlock the subtle energies within, revealing nature's innate power to heal."

Hahnemann's potentization process is believed to interact with water's unique ability to hold "memory." In homeopathic preparations, water acts as a medium that captures the energetic or vibrational "imprint" of the substance, which remains even at levels of dilution beyond molecular detectability. By imprinting this energetic "memory" into the water or alcohol medium, the remedy retains the healing essence of the original substance, resonating with the body's own energies.

This concept of water as a "memory holder" aligns with emerging theories that water's molecular structure can form organized patterns capable of storing information. Hahnemann's insight suggests that the memory of the original substance persists in the prepared remedy, interacting with the body's subtle energies to stimulate healing and balance.

Hahnemann's approach marks a revolutionary shift from material-based healing, proposing that life's subtle forces can be harnessed through precise preparation. His work continues to inspire holistic health perspectives, where water, with its capacity to store and transmit energetic information, plays a central role in healing. Hahnemann's legacy endures in the idea that the body, mind, and energetic systems are interconnected, and that nature's most delicate forces hold profound potential for restoring health and harmony.

CHAPTER 5

VIBRATION, LIGHT, EVOLUTION

JOHN STUART REID

THE CYMASCOPE AND WATER'S MEMORY

John Stuart Reid's groundbreaking studies in cymatics, the science of making sound visible, offer profound insights into the connection between sound, vibration, and water's structure. His invention, the CymaScope, is a revolutionary device that captures sound waves as visible geometric patterns on water's surface, illustrating how water responds to and retains the memory of vibration.

Reid's work reveals that vibrations organize water into harmonious patterns, a reflection of the deep connection between sound and nature. Through the CymaScope, sound becomes a "language of geometry and resonance," where each frequency leaves a visible imprint. For example, the sound of a woman's voice generates intricate, harmonious patterns, demonstrating how water can store vibratory information in its structure.

"Water reveals the unseen harmonies of life through cymatics, a language of geometry and resonance."

Reid highlights water's unique ability to act as a resonant medium, capturing the vibration of life itself. His work suggests that water is not just a passive element but a dynamic "canvas" that holds the memory of sound and vibration.

"Water holds memory in its patterns; every sound leaves a visible mark."

Through cymatics, Reid connects water to the vibratory origins of life, offering a scientific perspective on how water reacts to and organizes itself in response to frequency and vibration.

The CymaScope has opened new avenues for understanding and exploring the interplay of sound and water across numerous fields:

- Asteroseismology: Studying star vibrations.
- Biology & Botany: Understanding how sound influences cellular structures and plant growth.
- Hematology: Exploring how sound patterns affect blood flow.
- Musicology: Revealing the visual beauty of musical tones.
- Quantum Physics: Investigating vibration as the foundation of matter.
- Sound Therapy: Using vibration for healing and wellness.

Reid's innovation parallels the significance of the microscope and telescope, which expanded humanity's view of the micro and macro realms. The CymaScope opens a new frontier, making the invisible world of sound and vibration visible for both scientific study and artistic appreciation. By showing how water captures and reflects sound, Reid gives us a window into the vibratory essence of life.

This understanding of water as a resonant medium aligns with the idea that vibration underpins all matter in the universe. Reid's work with the CymaScope demonstrates that water doesn't just hold form–it holds memory, reflecting the unseen harmonies of life.

RICHARD MERRICK

INTERFERENCE THEORY AND HARMONIC EVOLUTION

Richard Merrick is a trailblazer in harmonic resonance, exploring how vibrational patterns shape life and the cosmos. His Harmonic Interference Theory (HIT) reveals how water's unique structure acts as a resonant matrix, transforming universal energy into the geometries of life—from DNA spirals to the symmetry of marine organisms.

Through his work, Merrick uncovers the deep interconnectedness of all things, showing how water bridges the energetic and physical realms. As he states: *"The spiral is the handwriting of the universe, a signature of harmonic resonance that binds all levels of existence into a coherent whole."*

Merrick's insights offer a profound understanding of life's harmonic blueprint and its cosmic origins.

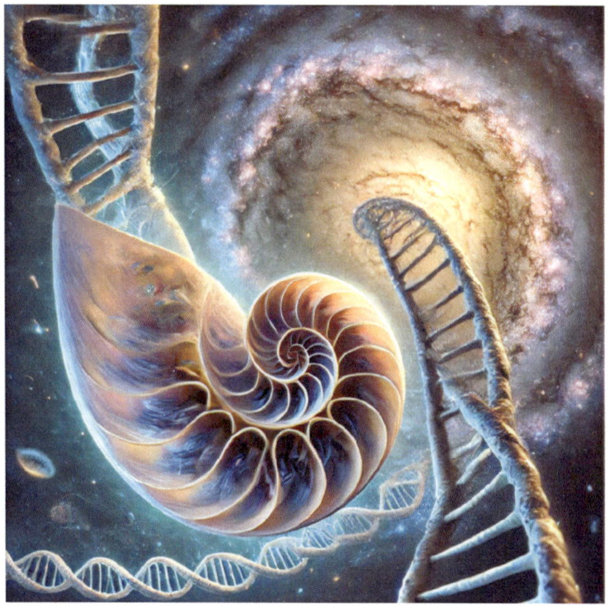

Water as the Universal Resonance Matrix

Water serves as a resonant matrix, mirroring the cosmos and capturing the vibrational essence of the universe within its quantum lattice. Its molecular structure enables standing waves–vibrations that reflect and amplify within enclosures such as cell walls, eggs, or wombs. These enclosures create the conditions necessary for harmonic patterns to form, shaping the geometries of life.

As Richard Merrick explains, water and Carbon-12 are the primary ingredients for life, yet the harmonic qualities of existence only emerge under pressure and inside protective environments. Whether within a spherical cell, a gravity well, or even the resonant chamber of a guitar string, life's interference patterns arise from these enclosed spaces, reflecting harmonic multiples of fundamental frequencies.

The Symphony of Life's Geometry

The interplay of electromagnetic and acoustic waves in water guides the formation of life's sacred geometries, embedding harmonic ratios into the physical structure of living beings.

- **The Spiral of DNA and Seashells**

DNA's double helix reflects the logarithmic spirals of seashells, such as the nautilus. These spirals are shaped by water's torsional forces and embody the harmonic ratios that define the universe's handwriting. Water's resonance translates these patterns into the building blocks of life, connecting molecular structures to cosmic geometries.

- **Starfish and Pentagonal Symmetry**

The fivefold symmetry of a starfish exemplifies harmonic ratios imprinted by water's interference patterns. This symmetry aligns with the golden ratio, demonstrating water's role in creating the recurring motifs found throughout nature.

- **Turtles and Fractal Patterns**

The tessellated shell of a turtle, with its intricate hexagonal and pentagonal designs, reflects fractal geometries formed by water's vibrational matrix. These fractals repeat at multiple scales, from molecular formations to organismal designs, illustrating theinterconnectedness of life's forms.

The Language of Vibrations

Water transforms universal energy into matter by harmonizing vibrations into organized forms. Each molecule resonates with universal frequencies, creating interference patterns that guide the assembly of cells, tissues, and organisms.

- **Water's Role in Morphogenesis**

During embryonic development, water's resonance interacts with genetic information to orchestrate the formation of cells and tissues. Merrick suggests that harmonic energy, channeled through water, shapes the growth and differentiation of living beings.

- **Harmonic Interference and Enclosure**

Life's harmonic resonance grows outward from a beating heart or central point, but it is attenuated or damped at Fibonacci nodes. Smaller organisms, like starfish, exhibit undamped, regular geometries, while larger animals develop proportional nodes that create joints and spaces between bones.

Merrick explains this dynamic through the blending of outward harmonic series with inward Fibonacci spirals, forming a first-derivative spatial Gaussian. When plotted in polar coordinates, these interference patterns produce heart-like or apple-like shapes, repeated across scales in living organisms.

Resonance and the Cosmic Song

Life's diversity arises from the interplay of vibrational patterns harmonized by water. Merrick's Harmonic Interference Theory reveals that these patterns mirror the structure of galaxies and the movement of stars, connecting the microcosm of living beings with the macrocosm of the universe.

- **Harmonics in Music and Nature**

Just as harmonic interference shapes life's forms, it also defines our perception of music. The human auditory system recognizes 12-step harmonies because it is attuned to the same interference patterns that guide life's architecture. This resonance binds us to the universal rhythm of existence.

- **The Sacred Geometry of Water**

Water's resonance creates sacred geometries, including the logarithmic spirals of DNA, the radial symmetry of marine life, and the fractal designs of turtle shells. These geometries reflect the harmonic principles encoded in water's vibrational patterns.

The Heart of Life

At the core of Merrick's Harmonic Interference Theory lies the beating heart—a central resonance point from which life's geometries emerge. The Fibonacci series spirals inward, while the harmonic series expands outward, creating a symphony of balance that reflects the essence of life.

This dynamic interplay forms the foundation for life's architecture, from the precise geometries of smaller organisms to the proportional complexity of larger beings.

- **The Cosmic Interconnection**

Water connects all forms of life to the universal lattice, harmonizing vibrations across scales. From the spiraling pathways of DNA to the tessellated patterns of a turtle's shell, Merrick's insights reveal the universal resonance carried through water, weaving a symphony of interconnectedness. As Merrick states: "*The shapes of life are not mere accidents of evolution but the echoes of a universal song, carried through water and resonating in every living form.*"

This narrative bridges Merrick's scientific insights with the beauty of nature's geometry, revealing water as the harmonic mediator of life and the cosmos.

LIGHT, GRAPHENE, TIME

THE BODY AS A LIVING HOLOGRAPH OF LIGHT AND SOUND

"The hourglass reminds us that each grain of sand is a fleeting moment, precious and irreversible, cascading into the reservoir of eternity."

Light, graphene, and time share a profound structural similarities;–they each mimic the form of the hourglass. This resemblance is no coincidence but a striking alignment of cosmic design, revealing their interconnected roles in the fabric of existence.

Light flows as waves, converging and diverging in cycles, much like the narrowing and expanding shape of an hourglass. Its energy channels through space and time, creating a rhythm that mirrors the dual flow of the hourglass.

Graphene, with its helical lattice and hexagonal precision, embodies balance and symmetry. Its molecular structure echoes the hourglass's form, suggesting a coordination of natural design that bridges light and time at the molecular level.

Time, visualized as the flow of sand (molecules) in an hourglass, captures the essence of movement and transformation. Each grain represents moments passing, converging at the narrow center and expanding outward, showing the assembley of cycles and evolution.

These parallels are not mere coincidences but coordinates within a universal blueprint, a resonance of form and function that unites light, graphene, and time. Together, they reflect a deeper order–a cosmic choreography where structure, energy, and transformation are all intricately woven into the hourglass of existence.

EDDIES AND VORTEXES

LIFE WITHOUT END

"Matter can neither be created nor destroyed; it only changes form," mused Walter Russell, hinting at the vast ocean of aether and water that holds all existence.

Through this cosmic sea, life flows in endless eddies and spirals, ever-changing yet never truly gone. Each atom dances to a universal heartbeat, expanding and contracting, breathing with the pulse of creation.

"Energy cannot be created or destroyed; it can only be changed from one form to another," wrote Einstein, echoing the ceaseless journey of life.

We are swept along, sometimes caught in still eddies of time, spiraling within, held in moments of quiet readiness before flowing outward once more. Like water, life merges and separates, yet always flows, seeking new forms in a cycle without end.

As Revelation reminds us, *"I am the Alpha and the Omega," says the Lord God, "who is and who was and who is to come, the Almighty."*

Life, bound by the aether, transcends beginning and end, forming a continuous loop of rest and renewal. "We do not die; we but change in form," Russell assures us, for the soul moves onward, part of the eternal. In this dance, life rises and falls like the ocean's waves, an infinite journey, spiraling us deeper into the mystery of becoming.

AS ABOVE, SO BELOW

This loop, a timeless exchange between earthly and heavenly realms, is the essence of life's continuity. We, like the atoms within us, are part of a flow that recycles and renews in a ceaseless cycle. It is a journey without beginning or end, a timeless dance where life is continually reborn and restored, moving between realms in an eternal cycle of rest, renewal, and regeneration. In this way, we, too, are eternally alive within the infinite loop of existence, cycling between heaven and earth, woven into the spiraling network in a hall mirrors. Where you are looped is where you are hooped.

"The hourglass teaches us that time is both finite and infinite: the sand falls, yet the cycle can begin gain"

THE LIFE-WAVE OF CREATION

STANDING WAVE, THE UNIFYING FORCE

"The universe, like life itself, thrives in the interplay of opposites: light and dark, creation and destruction, order and chaos." – Inspired by the concept of Yin and Yang

The standing wave can be seen as a microcosmic reflection of the larger cosmic principles, such as Yin and Yang, the formation of galaxies, and the primordial state described in ancient texts. It is a beautiful reflection of how opposing forces interact to create balance, order, and life itself. Just as galaxies condense from swirling gas, life forms from the interaction of energy and matter—brought together in a dance of expansion and contraction.

Water, energy, light, and matter are all expressions of the standing wave, this unifying force that flows through all things. From the formation of stars to the beating of our hearts, it is this wave that orchestrates the symphony of creation, continuously cycling between chaos and order.

In the ancient view, Sanskrit's Cosmic Egg represents the primordial gas from which all forms emerge, much like the standing wave that condenses pure energy into physical matter. This gas, an undifferentiated substance, holds the potential of all creation. It is in the interaction of forces, like gravity and centrifugal motion, that matter condenses and structures into the tangible world we experience. This mirrors the creation of galaxies and stars from swirling gas—a cosmic dance. that continues to shape the universe.

The standing wave is more than just a physical phenomenon. It is a cosmic principle that governs the birth of worlds and the flow of life. It creates the containers within which life takes form, offering both structure and freedom. It is in this dance of opposing forces that the beauty of existence unfolds, both in the stars above and in the cells within.

"In the beginning, there was chaos. And from chaos, all that is known was born, woven from the fabric of opposing forces, and through them, balance is achieved."– Inspired by ancient creation myths

WATER AND THE COSMIC DANCE OF OPPOSITES

Walter Russell's Sex Principle of Life illustrates the universal law of creation: the dynamic interplay of opposites. He described it as the division of equilibrium into male and female expressions, their merging creating the rhythm of life itself. Water, embodying this principle, becomes a living example of the cosmic dance. In its dual states–flowing as liquid and transforming into vapor or ice–it mirrors the eternal rhythm of giving and regiving.

The unification of opposites in water reflects the creative force of nature. Male and female energies–symbolized in water's polarity and its role as a conductor of life's currents–intertwine in a harmonic flow. This cosmic dance, like the merging of rivers into oceans or the interplay of rain and evaporation, sustains life, illustrating the unity of creation through the balance of opposites.

"Sex is the division of universal balance into opposites, their reunion multiplying life. This eternal dance of giving and regiving drives all creation." - Walter Russell, *The Universal One and The Secret of Light*

"In the wave lies the secret of creation" - Walter Russell

Through the merging of opposites, the universe sings its eternal song of life—forever in balance, forever in motion."
– Inspired by the cosmic dance of creation

THE ETERNAL RHYTHM OF FLOW

WAVES, CYCLES, AND LIVING WATER

In every corner of the universe, there is a rhythm—a pulse that carries the essence of life through cycles and currents. Hermes spoke of this in his timeless wisdom, describing how "*everything flows, out and in; everything has its tides; all things rise and fall.*" This cosmic pendulum swings between opposites, moving from light to shadow, from stillness to motion. Every crest is mirrored by a trough, each wave a part of the eternal dance that balances the universe.

Dale Pond echoed this understanding, describing how *"every object in the universe is subject to multiple cycles of varying wavelengths and amplitudes."*

Life itself flows within these cycles, responding to rhythms that transcend time. From the macrocosmic waves of galaxies to the subtle currents within our cells, these cycles shape the universe, intertwining every being in a symphony of movement.

Viktor Schauberger saw water as a living substance, attuned to this universal rhythm. In its natural state, water spirals, flows, and shifts with a consciousness that aligns with the pulse of nature. Each stream and current reflects the principle of rhythm, embodying the dance of creation. Water's flow is not merely mechanical; it is alive, harmonizing with the cycles of life, a vessel for the universe's heartbeat.

Together, these perspectives reveal a profound truth: everything is connected through waves and cycles. Water, with its graceful flow, mirrors the balance between light and shadow, expansion and contraction. It is the vessel of rhythm, a carrier of life's currents, flowing eternally in harmony with the cosmos.

Capturing the rhythmic flow from the coastline to the moon, demonstrates the interconnected cycles of nature weaving in a balancing act left to right.

BRIDGE BETWEEN SPIRITUAL AND PHYSICAL REALMS

FRACTAL CONTINUUM LIES WATER

As Above, So Below", this universal design, evident in the golden ratio, spiral galaxies, and human DNA, reveals rhythmic, self-replicating shapes that bind the microcosmic and macrocosmic realms through a harmonious symmetry.

At the heart of this fractal continuum lies water, which not only sustains life but serves as the universe's "conductor of life." Water, with its unique hydrogen bonds, carries vibrational frequencies and electrical signals that connect the vastness of the cosmos to the smallest cellular forms, bridging scales and amplifying life's resonance. As energy flows through water, it generates fractal-like structures seen in river networks, root systems, cloud formations, and even in the intricate branching patterns of the human nervous system.

Science now reveals that fractals, self-repeating patterns embedded in nature, can be found in the veins of leaves, branching rivers, and the strikes of lightning. Water enables these forms, transferring charge and information throughout all living systems. Just as a river's flow reflects the greater ecosystem, each cell within our bodies holds the blueprint of the cosmos, connecting the cycles of existence with fractal precision.

This dynamic interplay positions water as more than a physical substance–it becomes a cosmic conductor, a medium through which all forms are woven into one vast tapestry. Every droplet in a river, like each cell in our bodies, contains a reflection of the whole. Water's role as a conductor, the brains, of energy and life connects us to the universal rhythm and reminds us of the unity that permeates all existence–from the heavens to Earth and from the cosmos to the cellular.

In a sense, fractal structures, like dendrites in the human brain, mirror the nervous system of the cosmos. Echoing Tesla's view of the human body as a receiver, these fractal forms position us within the electric impulses of the universe, tuning us to its energy in a profound and eternal harmony.

WALTER RUSSELL

ELECTRIC SEAS'

Walter Russell's philosophy extends the understanding of existence beyond the material world into the realms of energy, vibration, and consciousness. His illuminations revealed a the universe as a vast, interconnected field of energy—an "electric sea" where everything is both a receiver and a transmitter of electric waves. In his teachings, Russell emphasized that all living beings are inherently electric in nature, drawing energy from and contributing energy back into the cosmos in a perpetual exchange that sustains life and creation (like the auroras).

The secrets of existence, Russell taught, lie within the wave. The wave carries the essence of creation—an oscillating electrical rhythm that flows like the ocean, embodying cycles that echo throughout the universe. Just as the tides of the sea rise and fall in harmony with the cycles of the Sun and Moon, so too do we, as beings of light, respond to these cosmic rhythms —manifested as electrical currents and circuits. Our circadian cycles are reflections of a greater symphony unfolding in the electric universe—a symphony composed of waves that govern light, energy, and time.

Russell believed that the spectrum of the light wave, combined with the medium of aether and water, generates the vibrational dance that creates life. From this interplay, the spectrum of light unfurls, bringing forth not only the visual brilliance of the universe but also the chemistry and sound that build life's architecture. Within these vibrations, bio-geometry emerges as the foundational design of creation, a language spoken by the wave to manifest life from light itself.

"We are all electric creatures floating in the electric sea of this electric universe," Russell wrote, emphasizing that life's essence is not static but dynamic, charged with the vibrancy of cosmic motion.

This energy exchange flows from the universe into each of us, filling us with the power to create, to love, and to evolve, and it flows back from us to the universe, contributing to the collective consciousness that binds all creation.

Through understanding this, we come to see ourselves not as isolated beings but as resonant expressions within the universal wave. Our existence, like the wave, is woven into cycles of light, sound, and life, resonating with the harmony of a cosmic design that reflects the oneness of all.

"We are all electric creatures floating in the electric sea of this electric universe. Every electric thing in the entire universe breathes electrically into itself from the rest of the electric universe which is outside of itself. It also breathes out of itself into the rest of the universe." ~ Walter Russell, *The Message of the Divine Iliad*

BLUEPRINT: WEB OF LIFE

FLOWER OF LIFE MIRRORED ON OUR LANDSCAPE

The Electric Sea flows unbroken, weaving the geometry of life into a tapestry that transcends borders and cultures.

Nature uses only the longest threads to weave her patterns, so that each small piece of her fabric reveals the organization of the entire tapestry." ~ Richard Feynman

"Electricity is the universe's heartbeat, connecting all that is, seen and unseen." ~ Walter Russell

Reflection is the Web of life, Flower of Life, which its intricate geometry appears like a cosmic blueprint woven through water's crystalline structure. Each curve, each line, each sacred (sacred means to bond) intersection is reflected upon the water's surface, and mirrored within, is the fabric of creation encoded in water's molecular bonds. Water's unique structure enables us to reflect on this ancient wisdom which amplifiesthe geometric harmony of creation.

In its purest essence, water becomes the canvas upon which light inscribes its signature. Its molecular structure, fluid and adaptable, reflects the same energetic geometry that guides the universe. Water's network of bonds behaves like an electrical system, conducting currents that pulse through its structure in discharge patterns. These patterns echo the same branching, fractal forms we observe in lightning, in roots, and in the cosmic web above. Water is thus the medium through which life's light—its very essence—is channeled and sustained.

This ability to mirror and conduct energy makes water the ideal conduit for the Electric Sea's flow. As light enters, it threads through the molecules, creating a dance of reflection and refraction. Each drop holds a thread of light, forming a network that channels energy through its bonds, connecting each molecule into a greater whole. The discharge patterns within water mirror the Flower of Life's sacred geometry, the universal blueprint. It is as if each bond and molecule within water carries the "imprint" of the cosmos, a pattern that threads light and life into the fabric of existence.

CURRENTS OF LIGHT BODIES

"The Flower of Life is the blueprint of creation and the genesis of all form."— Drunvalo Melchizedek

"Geometry is the archetype of the beauty of the world."— Johannes Kepler

Water's reflective nature also makes it the perfect mirror for the interplay of light and geometry, capturing the threads of light as they weave through its structure. The crystalline matrix of water holds these light threads, projecting them outward, creating a feedback loop between light, energy, and matter. Through this dynamic, water becomes not only a source but also a guide, reflecting the pattern of creation back into itself, allowing the Electric Sea to channel through it like an endless stream of life-giving force.

In every body of water–oceans, rivers, even the fluid within us–this electric and reflective property continues, binding us to the natural world and the cosmos beyond. The threads of light within water connect us to the heavens, capturing and amplifying the universal energy that sustains life in all its forms. This connection is water's gift: to reflect, amplify, and sustain the threads of light that weave through all of creation, forming an unbroken network that pulses with the energy of the cosmos.

Water is, therefore, both mirror and medium, a vessel that threads the light of the Electric Sea into every living thing, bridging the vast expanse of existence with the microcosm of life itself. It is in this timeless dance–where water reflects the Flower of Life and channels light's patterns–that we witness the interconnectedness of all things, a sacred thread that flows eternally within and around us.

ELECTRICAL DISCHARGE AND THE LIVING WATER OF CREATION

THE DIVINE GEOMETOR

The "Geometor," the cosmic architect, brings forth creation through sacred geometry and the resonance of water. The Flower of Life is not a static design–it evolves, energized by the electrical discharge of life and held together by the cohesive properties of water. This dynamic interplay of energy and water manifests the embryonic stages of existence, much like the intricate sketches of Leonardo da Vinci, who saw geometry as the language of creation. Da Vinci's studies of water's flow and its spiraling vortices echo the essence of the Geometor's work–revealing the unseen patterns that shape life. His understanding of the interconnectedness of art, science, and nature mirrors the cosmic dance of energy and form, where the sacred geometry of the universe is both blueprint and living art.

Water, the essence of life, is uniquely designed to capture and resonate with the vibrational memory of creation. Within its liquid lattice, the Flower of Life emerges as both an archetype and a dynamic biological process:

Cosmic Flow: Across the vast expanse of the cosmos, hydrogen and oxygen converge to form water–a fundamental medium through which electrical forces sculpt stars, galaxies, and the very fabric of the universe.

Earthly Reflection: On Earth, water mirrors the principles of the Flower of Life through its hexagonal crystalline patterns, a resonance that extends from the sacred geometry of snowflakes to the molecular and cellular networks within living organisms.

FLOWERS LIFE AND GRAPHINE

From Waves to Webs: The Flower of Life in Water

- **Electrical Discharge:** The formation of the Flower of Life is powered by an electrical pulse, like ripples on a pond. Water translates these pulses into vibrant patterns, embodying the wave signatures of creation.

- **Molecular Dance:** In water, the geometry of the Flower of Life appears in its hydrogen bonds. These bonds enable water to align with light, sound, and matter, acting as a bridge between the cosmic wave and the biological blueprint.

- **Hydrated Blueprint:** The development of life, from the cosmos to the human embryo, requires water to transport nutrients, energy, and information. Its molecular resonance amplifies the Flower of Life's vibratory essence, guiding the evolution of life.

Water and the Embryonic Stages of Creation

- **The Seed of Life:** The Flower of Life begins with a single drop—a vibration on the surface of water. This drop divides and multiplies, echoing the first moments of embryogenesis and the birth of galaxies.

- **The Egg of Life:** In both water and cells, the Flower of Life's design emerges as a sphere of life-giving potential. Water's hexagonal memory mirrors the pattern of DNA, the code of life.

- **The River of Life:** Just as rivers carve the surface of the Earth, water channels energy through the Flower of Life, shaping the evolution of creation in harmony with time and space.

"These structures are derived from the fact that a charged particle flow (or current) produces a ring of magnetic fields around itself, 'pinching' plasma into multi-filamentary strands, as can be seen on both cosmic and more localized scales."- Wal Thornhill, Electric Universe Model

ATOMIC RESONANCE AS A FRAMEWORK FOR LIFE

RICHARD MERRICK

All life on Earth is composed mostly of carbon-12 and water. This is the case because carbon-12 bonds or resonates with more simple elements than any other element in the universe. It is for this very reason that carbon-12 is the international standard for atomic weight and all other elements are measured against it. With 6 protons + 6 neutrons in its nucleus and 6 electrons orbiting in two shells, carbon-12 exhibits the lowest possible energy of all the elements and is said to be 'unbound', thereby creating the most stable atomic geometry possible. When mixed with water, carbon-12 creates endless chains of sticky amino acids capable of crystallizing into life.

This idea of life as a crystallization process is a good one because just as minerals align under pressure into lattices, coils of amino acids fold under pressure into three-dimensional protein structures, aligning into the familiar helical lattice of DNA. It is the geometric pressure of hydrogen atoms in water that helps create the lattice and give DNA its twist.

The water lattice begins as 4-fold tetrahedral units of 14 water molecules, aligning into 20 clusters to create the geometry of a 280-molecule water icosahedron. This structure then assumes a variety of stable, geometric sub-structures (such as its complementary dodecahedron) that form into even larger superclusters. At this mesoscopic scale of water, molecules arrange themselves into a 2-dimensional connectivity map of a regular 5- fold pentagon.

BIO-GEOMETRY: THE HEXAGONAL BLUEPRINT OF CREATION

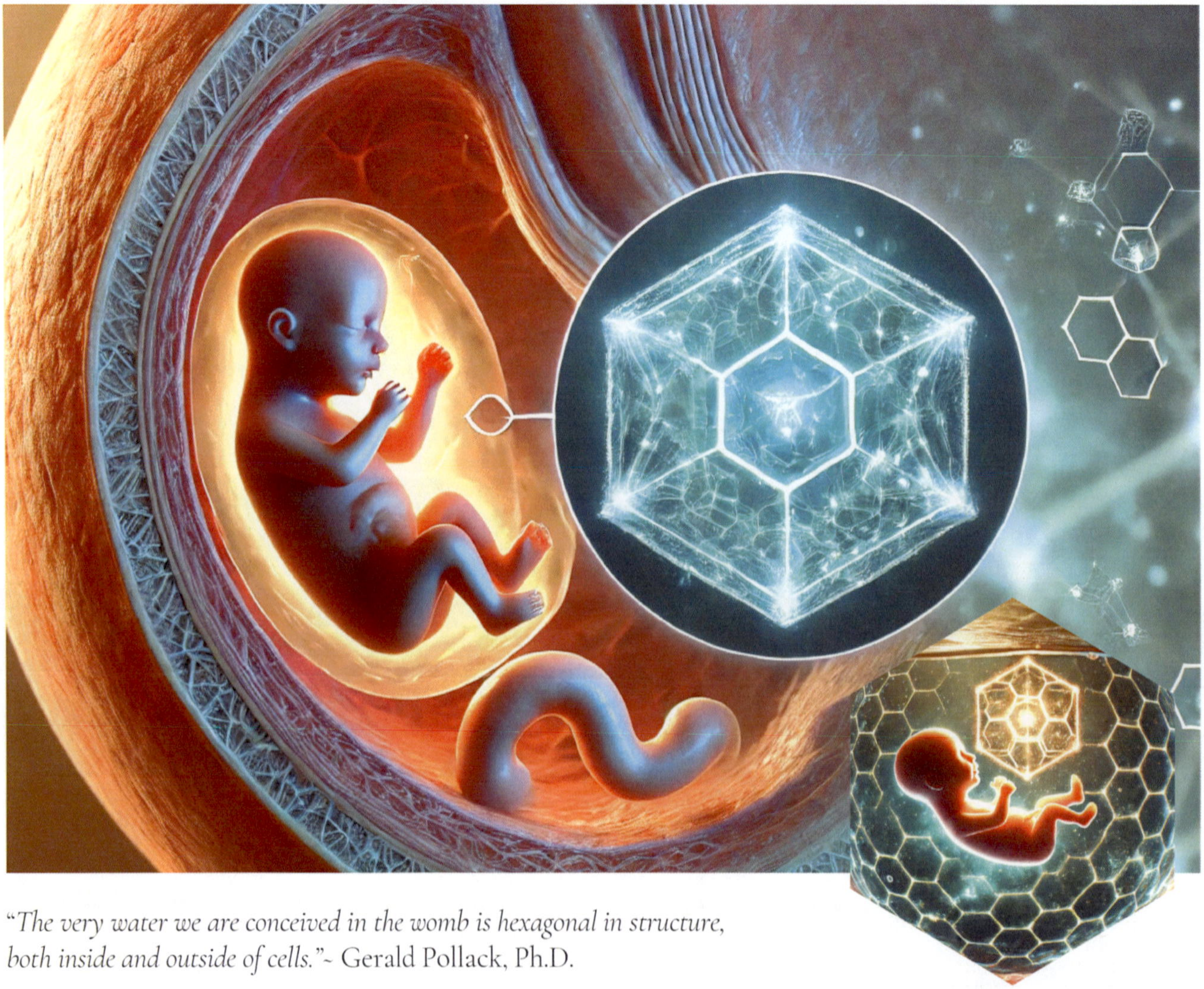

"The very water we are conceived in the womb is hexagonal in structure, both inside and outside of cells."~ Gerald Pollack, Ph.D.

Water's hexagonal structure represents one of nature's most efficient and conductive designs. In structured water (Exclusion Zone water), hexagonal layers form a natural electrical circuit within the body, separating positive and negative charges. This separation creates an energy gradient that powers cellular functions, akin to the way batteries store and release energy.

Hexagons: A Blueprint for Energy Transfer

The hexagonal geometry found in structured water optimizes energy transfer through its symmetry and organization:

- **Charge Flow:** Hexagonal water channels electrical charges efficiently, facilitating the flow of energy and light that powers cellular processes.
- **Self-Sustaining System:** This design allows for a continuous exchange of energy, creating a dynamic and regenerative system that mirrors the balance found in the cosmos.
- **Resonance and Coherence:** Hexagons act as a stabilizing framework for energy, ensuring that biological and energetic processes remain synchronized.

A Universal Conductor

The hexagonal structure is not confined to biology—it is a universal design that bridges form and function:

- **Crystalline Symmetry:** Found in snowflakes, honeycombs, and planetary formations, the hexagon reflects nature's pursuit of efficiency and coherence.
- **Molecular Bonds:** In organic chemistry, the hexagon provides stability and flexibility, enabling molecules to store and transfer energy.
- **Cosmic Design:** From the shape of Saturn's polar storms to the geometry of galactic formations, hexagons signify a universal language of creation that connects the microcosmic with the macrocosmic.

Bio-Geometry and Universal Intelligence

Bio-geometry aligns with this hexagonal intelligence, showcasing how nature uses this design to bridge physical and energetic realms. It transforms water into a conductor of life's energy, linking the molecular dance within cells to the rhythmic patterns of the universe. This shared geometry suggests a profound interconnectedness, where the same principles that govern energy transfer in cells are echoed in the structure of galaxies.

The hexagon, as both a pattern and a process, reveals the universe's inherent intelligence—an electrical design that sustains life, powers creation, and bridges the seen and unseen realms of existence.

The hexagonal structure embodying both the precision of nature's engineering and the nurturing essence of a mother's womb. The honeycomb, the pinnacle of geometric efficiency, maximizes storage and minimizes material use, sustaining an entire hive with its life-giving design.

The bee, a pollinator vital to ecosystems, symbolizes the intricate balance of energy transfer essential for life. Through its labor, the bee supports biodiversity, ensuring the propagation of plants and the sustenance of countless species.

Surrounding this scene, gentle waves of energy radiate like the electromagnetic pulses that power the cellular communication sustaining life. The honeycomb, like a microcosm of the cosmos, demonstrates how nature's designs are not random but rooted in universal principles—efficient, interconnected, and profoundly life-giving.

CHAPTER 6

VITAL FORCE'S TO TECHNOLOGY

WATER AS A CONDUCTOR OF ENERGY

Nikola Tesla's saw water as a conductor of these forces., It embodies many of Tesla's ideas, acting as a medium through which we can experience the interconnectedness of all life. Tesla's vision complements the scientific and spiritual perspectives on water, suggesting that by understanding water's nature, we get closer to the mysteries Tesla devoted his life to exploring.

Tesla's vision, with its focus on energy, frequency, and vibration, complements both scientific and spiritual perspectives on water, positioning it as a profound connector between physical matter and higher consciousness. Tesla famously stated,

"If you want to find the secrets of the universe, think in terms of energy, frequency, and vibration."

This approach aligns deeply with water's unique ability to absorb, transmit, and reflect vibrational information, making it an ideal medium for connecting physical life with the metaphysical.

In Tesla's view:

- **Water as a Conductor of Energy:** Tesla understood that water is one of the best conductors of electricity and energy, especially in its structured or pure form. This aligns with scientific findings that structured water has unique molecular arrangements that make it more conductive and bioavailable. The energetic properties of water are, therefore, amplified in its natural or structured state, supporting life processes at a cellular level.

- **Vibration and Resonance:** Tesla's theories suggest that all matter resonates at specific frequencies, and water, as a responsive medium, has a special ability to retain these vibrations. From the spiritual perspective, this idea extends to water's capacity to store and carry the essence of intentions, thoughts, and prayers, which Dr. Masaru Emoto explored in his studies on water crystallization. In the spiritual realm, water's ability to resonate and "remember" aligns with ancient beliefs that water is a carrier of consciousness.

- **Water as a Gateway to Higher Consciousness:** Tesla's concept of the universe as a vast network of vibrating energy fields implies that water, being a highly resonant substance, can help bridge the physical and non-physical. As a dynamic entity with memory and sensitivity to frequency, water acts as a medium that connects our physical being to higher planes of awareness, similar to the role it plays in both structured water science and spiritual practices.

- **Healing and Transformation:** Tesla's vision of non-physical energies contributing to healing aligns with water's role in vibrational medicine, where water is structured or energized to enhance health. Water holds the capacity to conduct and transmit healing frequencies, bringing about harmony within the body and encouraging balance, which echoes Schauberger's ideas of water as a life force.

In essence, Tesla's vision enriches our understanding of water as not just a physical necessity but as an active, conscious participant in life and evolution, capable of carrying and amplifying the energies of health, consciousness, and the universe itself. His ideas help bridge science, spirituality, and the ancient wisdom of water, positioning it as both a scientific marvel and a profound symbol of universal interconnectedness.

Viktor Schauberger, an Austrian forester and naturalist, is celebrated for his innovative insights into the nature of water, energy, and their connection to ecological balance. Schauberger saw water as a living entity with unique energetic properties, believing that water's natural flow and movement are essential for maintaining its vitality and life-giving properties. He observed that water, when allowed to spiral and flow naturally, maintains a "life force" that supports ecosystems, human health, and even technological innovation.

"Observe and copy nature."

VIKTOR SCHAUBERGER

WATER AS A LIVING ENERGY AND NATURE'S LIFE FORCE

"This simple yet profound statement summarizes Schauberger's philosophy. He believed that nature's methods, particularly water's natural flow and movement, are optimal and should guide our own systems. To him, water in its natural state is a source of life-giving energy that humans should respect and replicate in their designs.

"Water wants to spiral."

Schauberger observed that water naturally moves in spiral or vortex patterns, especially in rivers, which he believed was essential for maintaining its energy and purity. This natural spiraling motion allows water to oxygenate, cleanse itself, and accumulate vitality, which is disrupted by straight-line mechanical movement, such as traditional pumping.

"Whoever understands water will understand the universe."

Schauberger saw water as a microcosm of universal forces. He believed that by studying water's properties–its flow, structure, and energy–we can understand deeper cosmic laws, including balance, rhythm, and the cycles that sustain life.

"Water is the blood of the Earth."

This quote reflects Schauberger's view that water is to the Earth what blood is to the human body–a vital, life-sustaining force. He saw forests, rivers, and springs as part of a living system where water flows in natural cycles, nourishing plants, animals, and ecosystems.

"Living water is the foundation of all life, and when its vitality is disrupted, all life suffers."

Schauberger believed that when water is polluted or its flow is altered, it loses its "life force," impacting everything it touches. He observed that unnatural movement and pollution strip water of its vitality, which in turn affects the plants, animals, and humans who depend on it.

DR. KRISHNA MADAPPA

REVIVING LIFE THROUGH THE SACRED INTELLIGENCE OF WATER

Dr. Krishna Madappa has devoted over three decades to the study, transformation, and elevation of water–not merely as a substance, but as a living intelligence capable of restoring balance across all layers of life: body, mind, and spirit. As a pioneer in structured water and holistic health, his mission has always been clear: revive the planet by reviving the water.

The Sanskrit word "Jiva" means life–a reflection of his purpose to restore the life-force within water and, through it, reawaken vitality in both the human system and the Earth. His research and innovations have helped shape the emerging field of structured water technologies, including the widely respected JivaWater Devices, which use vortexing, frequency modulation, and oxygenation to re-energize water–making it more coherent, bioavailable, and profoundly life-giving.

Dr. Madappa sees structured water as a crystalline matrix, a dynamic molecular arrangement capable of holding and transmitting energy and information. This "living water," as he calls it, brings coherence to the body's internal environment, enabling cellular communication and supporting well-being across emotional, mental, and spiritual dimensions.

A Bridge Between Ancient Wisdom and Modern Science

Emerging research supports the idea that water can exist in a fourth phase–beyond solid, liquid, and gas. This "structured" or "exclusion zone" (EZ) water, as demonstrated by Dr. Gerald Pollack, forms near hydrophilic surfaces and displays unique physical properties, such as charge separation and energy storage. Structured water supports enhanced hydration, nutrient absorption, and metabolic efficiency. Its lattice-like geometry acts as a liquid crystal, suggesting water can store memory, amplify coherence, and resonate with life's frequencies.

"Water is a mirror and a messenger. It reflects who we are and offers us a chance to realign, regenerate, and remember the essence of our being."

This paradigm-shifting understanding is echoed in studies across biophysics, biofield science, and unified field theory, all showing that water's structure can be influenced by sound, light, and electromagnetic fields. Dr. Madappa's JivaWater Devices embrace these findings, aligning modern technology with nature's intelligent design.

Jiva Water – Re-Energizing Life at the Molecular Level

Jiva Water is not just structured–it is re-energized water with:

- Higher molecular density
- Lower surface tension
- Stronger hydrogen bonds

These qualities make it easier for your cells to absorb water, improving hydration and supporting a wide range of health benefits from enhanced energy to deeper cellular regeneration.
JivaWater Devices are designed using principles of biomimicry, replicating the natural vortex motion of rivers and streams to generate a piezoelectric effect–a gentle electrical charge that realigns and revitalizes the water's molecular structure.

Through this inspired fusion of ancient knowing and scientific precision, Dr. Madappa offers a profound key to personal and planetary wholeness. He reminds us that when we align with the intelligence of nature, we restore the flow of life itself.

PIEZOELECTRIC STRUCTURING OF WATER: SCIENCE ECHOES NATURE'S GENIUS

FEATURED RESEARCHERS: MARK JONES, MICHAELA DOH ALOVÁ, PETR KAČER | CZECH UNIVERSITY OF LIFE SCIENCES LEGACY

NATURES' INSIGHTS: VIKTOR SCHAUBERGER – NATURE'S VORTEX VISIONARY

Water is a shape-shifter, a living matrix, whose behavior is profoundly influenced by the structure of its hydrogen bonds. In this pioneering study, Mark Jones, Michaela Dohnalová, and Petr Kačer of the Czech University of Life Sciences examined how mechanical stress applied via the piezoelectric effect restructures water on a molecular level. Their research reveals that under such influence, water's hydrogen bonds shorten, leading to a more compact, cohesive, and bioavailable form of water.

This mirrors what Viktor Schauberger discovered through observing nature: as water flows in spirals and vortices, it becomes naturally structured, oxygenated, and energetically alive. Schauberger believed that nature's true movement is inward, implosive, and intelligent. This study shows that mechanical input via piezoelectric stress mimics those same principles, producing water that behaves in ways highly beneficial to both biology and industry.

The Heart of the Study: Shorter Bonds, Stronger Water

The research team explored how the piezoelectric effect (mechanical stress that generates electric fields) changes the length and strength of hydrogen bonds. Using tools like FTIR, NIR, NMR spectroscopy, pycnometry, and viscosity testing, they discovered:

- Shortened hydrogen bonds (1.7 → closer to 1.6 Å), making water more cohesive
- Increased molecular density
- Stronger surface tension and optimized viscosity
- Improved molecular organization and flow properties

Simulations confirmed that these effects support better hydration, faster nutrient transport, and enhanced solubility. In biological systems, this means water flows more efficiently through membranes like aquaporins, supports cellular uptake, and enables more efficient biochemical reactions.

Vortexing and JivaWater: Mirroring Nature

Viktor Schauberger's principles are now applied in modern technologies like JivaWater, which use vortexing, frequency modulation, and pressure differentials to restructure water. Devices like these are based on the same natural forces Schauberger observed and that this Czech study quantifies:

- Spiraling motion = mechanical energy
- Compression zones = piezoelectric charge induction
- Structured water = improved vitality and hydration

This validates that JivaWater mimics the mechanics of nature, restoring water to its optimal state.

Water responds to energy, form, and rhythm. Whether by natural vortex or engineered piezoelectric stress, we now have the tools to restore water to its intelligent, life-enhancing state.

Nature's Method vs. Scientific Replication

How Schauberger's Vortex Wisdom and Modern Piezoelectric Research Align

Aspect	Nature's Method (Viktor Schauberger)	Scientific Replication (Jones, Dohnalová, Kačer)
Water Movement	Spiraling vortex flow through mountain streams	Mechanical stress applied via piezoelectric materials
Energy Source	Natural implosion, pressure gradients, Earth's fields	Induced electrical charge from mechanical compression
Effect on Structure	Realigns water molecules, forming hexagonal clusters	Shortens hydrogen bonds, increases molecular density
Resulting Properties	Energized, oxygen-rich, highly absorbable water	Higher hydration potential, increased solvent ability
Measurement Tools	Observed through behavior of fish, forests, and flow	FTIR, NIR, NMR, molecular simulations, crystallography
Philosophy	"Understand and copy nature"	"Analyze and replicate molecular mechanics"
Application Legacy	Basis for modern structured water technologies	Validates design of vortex-based systems like JivaWater
Biological Relevance	Enhances forest vitality, water health, and soil	Improves hydration, nutrient transport, and cell support

BIOMIMICRY IDEAS

MIMICKING NATURE'S DESIGNS

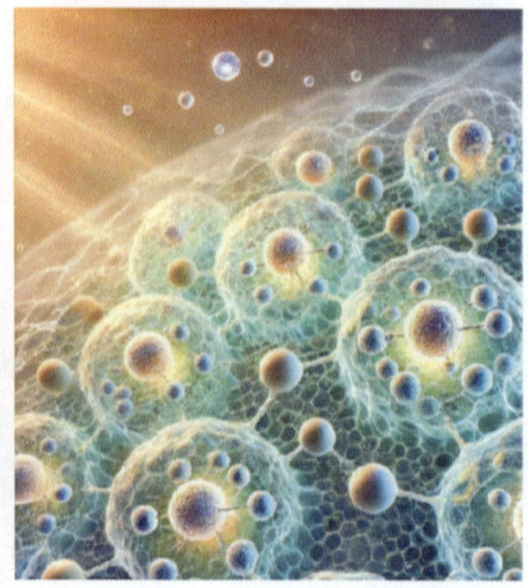

Mimicking Nature: Solar and Cellular Energy
Dr. Gerald Pollack's groundbreaking work on structured water (Exclusion Zone water) deepens this understanding. He discovered that in every cell, water organizes itself into layers that store and release charge, creating a gel-like matrix capable of absorbing light energy, particularly from sunlight and infrared radiation. "In every cell, water acts as a bridge between life and energy, its structure harnessing the sun's rays to power our biological functions."

The Role of Water in Energy Transfer
In biological systems, structured water serves as the medium for energy transfer, converting light into the electrical charges that fuel cellular processes. This parallels how water in technological systems stabilizes reactions and facilitates energy flow, such as in charging batteries. Both natural and human-made designs depend on water's unique properties to bridge the gap between energy sources and their applications.

Solar cells, human cells, and batteries all follow nature's timeless blueprint, demonstrating how the simplest designs–like water's structured lattice–create the most profound impact. Pollack's insights reveal that structured water acts as both a storage medium and an active participant in energy dynamics. As we innovate, we continually return to the principles nature has perfected over millennia, drawing from the same energy source that sustains all life: the sun. This interconnectedness underscores a profound truth: whether in cells or solar panels, the universe's designs are both functional and elegant, reminding us that nature's wisdom is a timeless source of inspiration. A Prophecy of Re-Creation The Hopi prophecy warns of humanity's journey to recreate itself through technology, only to rediscover what was known all along–the wisdom embedded in nature's design. As we marvel at the ingenuity of our creations, structured water humbly reminds us that the answers have always been flowing around and within us.

THE GEOMETRY OF PURITY

At the microscopic level, graphene oxide filters and aquaporin-inspired membranes perform a miracle—they purify water by allowing only the smallest molecules to pass through. Inspired by cellular membranes, these technologies mimic nature's selective filtration systems, removing contaminants with unprecedented efficiency.

In a world where clean water is scarce, these filters embody the elegance of natural design, showing how technology can meet humanity's greatest needs while maintaining harmony with the environment.

THE WISDOM OF TREES

Capillary Action and Biomimetic Pumps

Trees are silent engineers, lifting water from their roots to their highest leaves through the elegant mechanism of capillary action. Mimicking this natural process, biomimetic water pumps use surface tension and cohesion to distribute water without the need for external energy.

This innovation holds promise for remote regions, where efficient, low-energy water distribution systems can transform lives. Just as trees sustain entire ecosystems, these pumps remind us that life flows effortlessly when we align with nature's wisdom.

THE LOTUS EFFECT

Self-Cleaning Surfaces for a Cleaner World

The lotus leaf, revered for its purity, never holds onto dirt or water. Its microscopic structure creates a hydrophobic surface, inspiring coatings that repel water and prevent contamination.

These surfaces are revolutionizing water infrastructure, from pipelines to sanitation systems. By imitating the lotus, we learn to let go of excess and embrace simplicity, creating systems that are as resilient as they are efficient.

DR. WEST MARRIN

FRACTALS IN NATURE 'S DESIGN

D.L. Marrin holds a Ph.D. in water resources from the University of Arizona, along with B.S. and M.S. degrees in biological and environmental sciences from the University of California. A former Adjunct Professor at San Diego State University, he is the author of Universal Water, Altered Perceptions, and Hydromimicry, and has written extensively on water and biogeochemistry. Through Water Sciences & Insights, Marrin provides a forum for exploring diverse perspectives on water and supports various groups with environmental and educational initiatives.

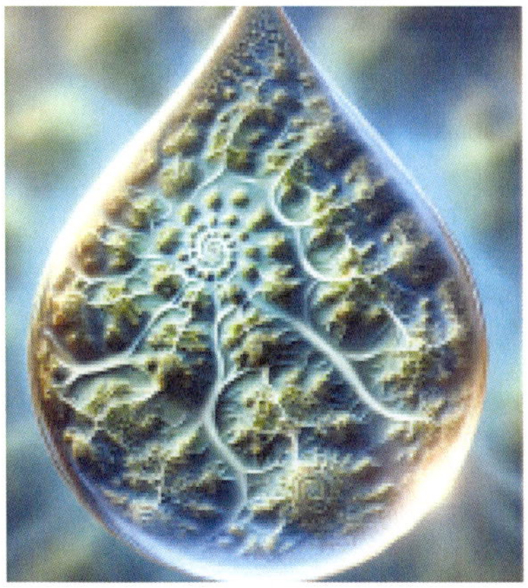

West Marrin, an applied scientist and founder of Water Sciences & Insights, has dedicated his career to examining the unique, self-repeating patterns present in natural watersheds and landscapes. These repeating geometric patterns, or fractals, appear at multiple scales in rivers, glaciers, and watersheds, forming a complex yet harmonious blueprint for water's role in shaping the Earth.

"A fractal is a repeating geometric pattern that has the exact same proportions on different spatial scales."

In watersheds, fractals create landscapes that might appear random or haphazard due to their absence of straight lines or perfect shapes. Yet, when viewed from a distance or up close, these organic patterns are consistent. Fractals in watersheds mean that large river systems viewed from an airplane reveal the same intricacy as a single stream segment observed close-up. Marrin's work highlights how these patterns directly influence a watershed's ability to maintain water quality, regulate flow, and buffer against extreme events, which is essential for ecological health.

"The intricate patterns created by water influence a landscape's hydrologic regimes, water quality, and resilience against natural events."

Although these fractal patterns have been largely overlooked in traditional human-made landscape design, recent interest in biomimicry has inspired applications of fractal designs in eco-compatible creations. Marrin's work explores how human landscapes, such as green roofs, artificial wetlands, and stormwater networks, can emulate natural fractals to support sustainable water management and ecosystem resilience.

"Fractal-like spatial patterns in watersheds inspire sustainable designs that enhance water management, environmental harmony, and resilience."

Through the concept of hydromimicry, Marrin advocates for emulating nature's fractal patterns to create landscapes that function in balance with the environment. His work illustrates how the intricacies of watersheds reveal deeper insights into water's role on Earth, offering valuable lessons for sustainable design and resilience in human systems.

"Water is the pulse of the planet, cycling through systems, nourishing ecosystems, and connecting continents."

"It has not escaped the notice of some biologists that the complex structure of the cytomatrix is fractal." — Dr. Mae-Wan Ho

GEORGE WISEMAN

THE REJUVENATING POWER OF HYDROGEN

Water, the foundation of life, consists of two parts hydrogen and one part oxygen–two gasses essential to health, rejuvenation, and the very essence of existence. The role of hydrogen, as the universe's simplest and most abundant element, is far more profound than previously understood. We are 62% hydrogen by volume and virtually every bodily process requires hydrogen directly or indirectly. Having enough hydrogen plays a pivotal role in enhancing vitality and sustaining life.

As George Wiseman aptly states: *"Hydrogen is simple yet profound, the building block of the entire universe, and essential to nature and vitality in all life."*

Wiseman, a pioneering researcher in transformative hydrogen applications, has dedicated his life to uncovering the remarkable potential of hydrogen, not just for energy production but also for human wellness. His focus centers on Brown's Gas (HydrOxy)–a unique, water-derived gas that is revolutionizing both health and sustainable energy. Unlike its counterpart, OxyHydrogen, which lacks Electrically Expanded Water (ExW), HydrOxy incorporates this vital component, making it about 30% more therapeutically effective than OxyHydrogen.

Understanding Brown's Gas and Electrically Expanded Water (ExW)
HydrOxy is a remarkable advancement in hydrogen-based technologies. The inclusion of Electrically Expanded Water (ExW)–a negatively charged cold plasma form of water ($H2O^-$)–is what sets it apart from other hydrogen and oxygen gas mixtures. ExW not only enhances the efficacy of Brown's Gas but also significantly increases its therapeutic and transformative potential.

One of ExW's extraordinary properties is its ability to facilitate the production of EZ Gel, what many still refer to as "EZ Water." However, as Wiseman clarifies:
"EZ 'water' is not water. It's H3O2-, not H2O, and it's not a liquid—it's a gel. EZ Gel is likely one of the most important discoveries ever because it's WHY life can exist."

EZ Gel, first described by Dr. Gerald Pollack, is fundamental to the structure and function of living cells. The additional electrons provided by ExW allow for a dramatic increase in the production of this gel from water. This enhancement enables Brown's Gas to help the natural processes of detoxification, nutrient transport, cellular repair, cellular communication, energy transfer, and coherence across biological systems far more effectively than traditional hydrogen therapies.

The Mysteries of "New Water"

Wiseman's research has also uncovered another groundbreaking phenomenon: the creation of "New Water." When Brown's Gas is burned, the resulting steam condenses into water that is Deuterium-Deleted (nearly zero deuterium) and, astonishingly, devoid of any prior memory. This New Water is a blank slate, free from environmental or energetic imprints, making it uniquely responsive to its surroundings.

As Wiseman shares: *"New Water immediately starts accumulating the 'knowledge' and information from its environment. For example, when I placed a bottle of New Water in the fridge next to orange juice, the next day it tasted exactly like orange juice. Perhaps this is how Jesus turned water into wine?"*

New Water's capacity to absorb and transmute information makes it a powerful medium for homeopathy and DNA information transfer. This property could significantly enhance the effectiveness of experiments like those of the late Dr. Luc Montagnier, where water acted as a carrier of DNA signals. Such discoveries align with Wiseman's vision of hydrogen and water as mediums of transformation, healing, and interconnectedness.

The Intersection of Energy, Biology, and Creation

At its core, Wiseman's work emphasizes the balance of forces within water and hydrogen systems. The dynamics of centripetal implosion and centrifugal expansion within water reflect the natural law of balanced polarity–what Walter Russell referred to as the "sex principle" of creation. These forces mirror the yin and yang, the masculine and feminine, and the dynamic interplay that sustains life and the universe.

Wiseman explains: *"This balance of forces is critical for achieving homeostasis, not just in the human body but also for the planet itself. By understanding and harnessing these natural dynamics, we can restore harmony, health, and vitality."*

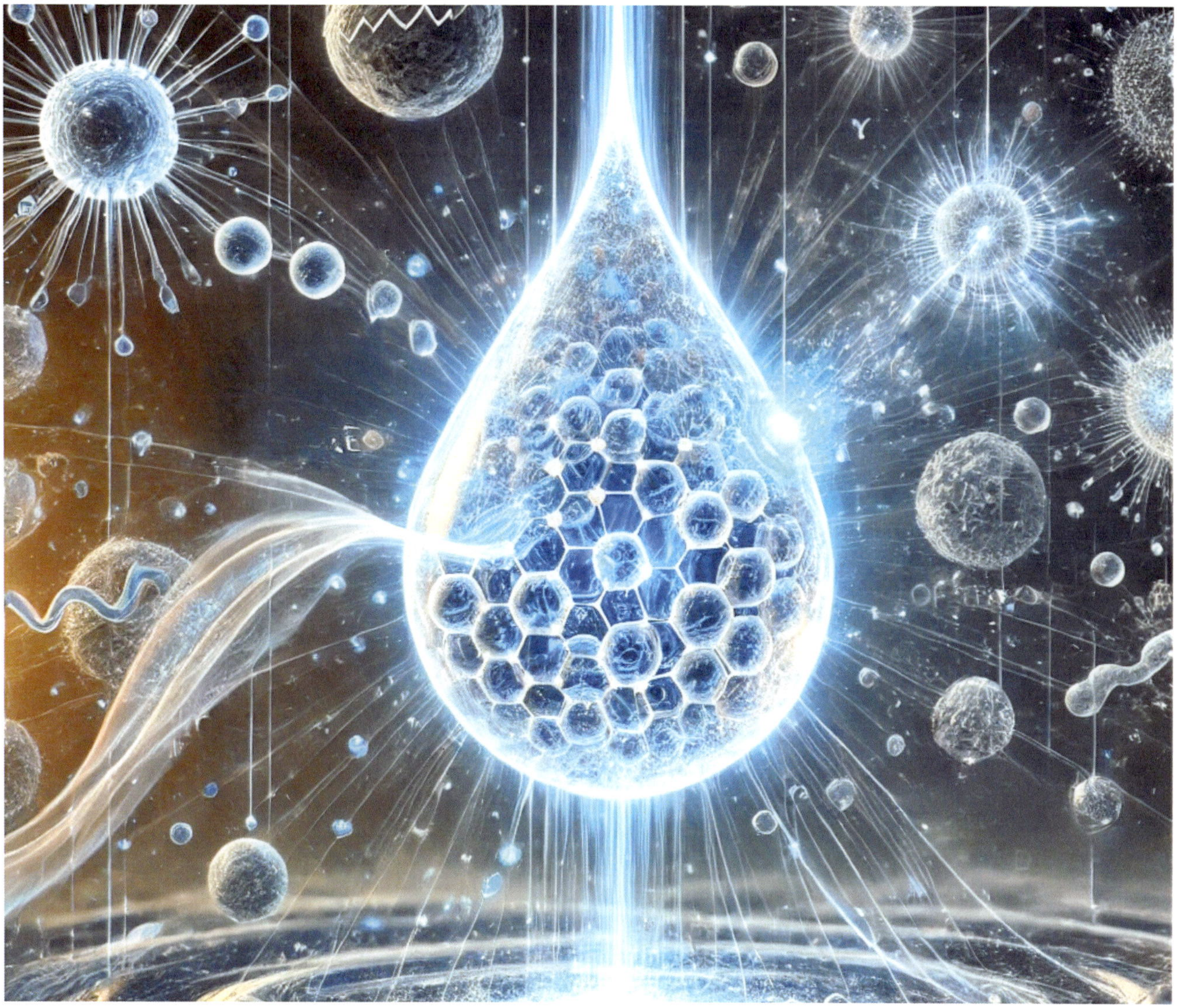

Hydrogen: The Element of Life and Energy

Hydrogen is not just a key element in water; it is the medium of life. Its simplicity belies its profound impact, as it bridges the gap between sustainability in nature and vitality in human life. Hydrogen-enriched water, particularly when infused with Brown's Gas, has the potential to:

- Enhance cellular repair and detoxification.
- Facilitate nutrient transport and energy production.
- Generate Deuterium-Deleted New Water, amplifying therapeutic and homeopathic applications.

Wiseman's AquaCure technology brings these possibilities within reach, enabling individuals to create HydrOxy-enriched water and tap into the transformative power of hydrogen from their own homes.

As Wiseman concludes:

"Water and hydrogen are not just resources–they are profound nutrients that embody the essence of creation and the potential for rejuvenation. By aligning with these principles, we unlock the key to energy solutions that reflect the harmony and balance of natural law."

WATER'S VITALITY

REVEAL THE DYNAMIC WAYS WATER SUPPORTS AND SUSTAINS LIFE

Water is more than a substance–it's a "living engine" vital to life on Earth and within the human body. Moving in spirals, vortices, and helical patterns, water naturally oxygenates, purifies, and conserves its energy, enabling it to sustain ecosystems and nurture the life-giving properties within every cell.

Enhanced Oxygenation and Vitality through Spiral Flow

Water's spiral motion, observed by Viktor Schauberger, allows it to incorporate air as it flows, increasing its oxygen content. This naturally oxygen-rich, "alive" water, often found in mountain streams, supplies essential oxygen to support thriving ecosystems, nourishing aquatic plants, fish, microorganisms, and even the cells in the human body.

Inspired by this principle, modern vortex-based water purification systems and spiral irrigation techniques mimic nature's oxygenating flow. These systems maintain water's vitality without chemicals or high-pressure mechanisms. Oxygenated water nurtures biodiversity and boosts soil health, providing nutrient-rich, oxygenated water to crops and delivering a natural vitality to all it sustains, from the tiniest microbes to the largest organisms.

Self-Purification through Vortex Motion

Water's vortex flow creates a natural cleansing action as centrifugal forces push impurities outward, allowing water to shed pollutants. This self-purification process is what keeps rivers and streams fresh, supporting ecosystems capable of sustaining life in balance.

By emulating this principle, vortex-based filtration systems offer low-energy, efficient solutions for water purification, reducing the need for chemical treatments and improving water quality in both urban and rural environments. These self-cleaning systems mimic nature's own methods, promoting water purity while protecting habitats from contaminants. By adopting this natural approach, we can design more sustainable, eco-friendly water management systems that reduce environmental impacts and promote wellness.

Energy Conservation and Reduced Friction in Water Flow

Schauberger observed that water's spiral movement reduces friction, conserving energy and allowing it to flow effortlessly. This "life force" preserved in flowing water prevents stagnation, resists pollution, and enhances its natural vitality.

Inspired by this efficient flow, vortex-based technologies and biomimetic designs in urban water management help reduce friction, conserving energy and optimizing natural flow. In cities, biomimetic stormwater systems with permeable surfaces, rain gardens, and bioswales channel water naturally, reducing erosion and preventing flooding. In agriculture, spiral irrigation systems deliver water more effectively, conserving energy and boosting crop resilience.

By honoring these natural dynamics within water, we foster the same vitality in our environment–and in our own bodies.

A broader view of spiraled channels in a farm landscape. This setup enhances oxygenation and vitality, bringing life to the surrounding plants and symbolizing the rejuvenating benefits of this nature-inspired irrigation method.

Most of our tap water arrives "flat-lined," having lost its natural vitality after flowing through rigid, straight pipes. By adding a vortexing system that mimics nature's design, we can re-oxygenate and energize the water, bringing it back to life. This "living" water enhances biological systems, providing the energy and balance found in nature.

WHIRLPOOLS OF ENERGY

THE WORKHORSE OF THE UNIVERSE

"Why have we ignored the vortex, the workhorse of the universe?" questioned William Baumgartner. The vortex, with its capacity to sustain energy while reducing friction and drag, is an essential engine of nature.

As Josef Tyls, Ph.D., observed, *"Nature uses vortices to reduce the energy required to perform work, minimizing both friction and drag; this is not just for the airways, but also the waterways."*

Viktor Schauberger went further, seeing the vortex as the universal machine through which all things take form. He noted, *"Every living creature, every physical form in the act of bringing forth its visible form out of its archetype idea, passes through the swirling vortex motion in order to manifest. How could we have missed this universal machine?"*

But the vortex's role is more than energetic efficiency; it is alchemy in motion. Through the swirling forces within water and air, the vortex disassociates impurities, cleanses, and transforms. Elements shed impurities as they spiral, embracing a purified state within the alchemical dance of creation and dissolution.

In these "whirlpools of energy, matter is cleansed, energized, and transfigured, embodying the pure essence of alchemy."

Schauberger viewed each vortex as a living, cleansing machine, capable of transforming energy back into vitality. From water currents to tornadoes, vortexes are not just phenomena but embodiments of nature's ability to continually purify, renew, and balance life.

"A vortex, much like a living stream, spirals with a concentrated stillness at its heart—a serene and shadowed center encircled by a radiant, whirling light. Just as nature's storms form a calm eye through counter-rotating winds, the eye of the vortex forms in balance with opposing forces, held in place by the flowing electric field of light. This movement, as Schauberger saw it, is nature's design: a dynamic dance of energy, where each side of the beam directs the light differently, creating a stable core amid perpetual motion."

Technical Applications: This insight is used in wind turbines with helical blades, aircraft design, and building structures that mimic the twist of tree trunks or shells.

Formation of Structured Water and Cellular Benefits

Schauberger observed that water flowing in spirals and vortices naturally organizes into a structured, coherent state. In this structured form, water molecules align into hexagonal clusters, creating "living" water that is more stable and bioavailable for organisms.

Modern Applications and Benefits

Structured water has been shown to improve hydration and nutrient absorption in living cells, benefiting plants, animals, and humans alike. Sustainable irrigation systems that promote natural water structuring help crops absorb nutrients more efficiently, enhancing agricultural resilience. For human health, structured water supports cellular hydration, metabolic efficiency, and overall vitality. This coherence is a cornerstone for biological wellness, contributing to sustainable health practices that harmonize with water's natural design.

Biomimetic Water Management and Eco-Friendly Design

Schauberger's philosophy, "observe and copy nature," emphasized that nature's designs–such as spiral river flows, leaf shapes, and natural filtration–provide efficient, self-sustaining models for technology. His work inspired biomimetic designs that echo these principles, creating systems that naturally conserve resources and support environmental health.

Biomimetic designs for urban water systems, inspired by Schauberger's work, include permeable pavements, green roofs, and rain gardens that capture, filter, and reuse water. These designs prevent urban flooding, reduce erosion, and support groundwater recharge. In agriculture, spiral-based irrigation mimics natural flow patterns, reducing water waste and promoting soil health. Schauberger's insights have helped create a foundation for sustainable design that harmonizes with ecosystems rather than disrupts them.

Implosive Hydropower for Renewable Energy

Schauberger proposed implosive hydropower as an alternative to conventional, energy-intensive methods. Unlike explosive systems that generate power through combustion, implosive energy systems utilize inward-spiraling motion to generate force efficiently and sustainably.

His vision of an implosive energy system relied on working with natural forces to reduce friction, conserve energy, and achieve efficiency. Hydrogen engines with water injection share these goals: they produce fewer emissions, conserve energy through optimized combustion, and can generate the power needed for heavy-duty applications without the waste associated with traditional fossil fuels. They are now in view.

Schauberger's Legacy: A Sustainable Path Forward

Through his understanding of water's behavior, Schauberger highlighted the immense potential of natural flow patterns to purify, energize, and sustain life. His ideas have inspired a wave of sustainable technologies that honor water's natural movements, supporting clean, efficient, and life-enhancing systems. By adopting Schauberger's approach of observing and mimicking nature, we can develop innovative solutions that reduce environmental impact, conserve resources, and promote balanced ecosystems.

THE LIFEBLOOD OF THE EARTH

SEAWATER'S ROLE IN SUSTAINING LIFE

Seawater's mineral-rich composition is the result of a profound interplay between the earth's landscapes and its oceans. Rainfall erodes minerals from mountains and streams, carrying them into rivers that eventually flow into the sea. This natural runoff enriches seawater with elements like magnesium, potassium, calcium, and iodine—nutrients essential for life. Over millennia, this cycle has created an oceanic reservoir of life-sustaining minerals, a resource not only for marine ecosystems but also for agriculture and human health.

When properly diluted, seawater can enrich depleted soils, stimulate microbial activity, and provide trace minerals often missing in conventional fertilizers. This infusion creates a fertile foundation for plants to thrive, yielding healthier crops and nutrient-dense food.

When properly diluted, seawater can enrich depleted soils, stimulate microbial activity, and provide trace minerals often missing in conventional fertilizers. This infusion creates a fertile foundation for plants to thrive, yielding healthier crops and nutrient-dense food.

Mimicking Nature's Cycles

In nature, the ocean's minerals are carried back to the land through rain, wind, and rivers, replenishing terrestrial ecosystems. Using seawater in agriculture mimics this ancient process, aligning with natural cycles to maintain soil fertility. This practice reflects the interconnectedness of all life and how the ocean's abundance sustains the earth's ecosystems.

A Symbiotic Opportunity: Desalination and Agriculture

Desalination, a process that transforms seawater into freshwater, produces mineral-rich brine as a by-product. Rather than discarding this concentrated solution, it can be repurposed to enhance soil health. When diluted to safe levels, this by-product offers a sustainable way to reintroduce essential minerals into agricultural systems. In this way, desalination not only addresses water scarcity but also supports regenerative farming practices.

Benefits for Soil and Plants

1. **Enhanced Soil Structure:** Minerals like calcium improve soil texture, increasing its capacity to hold water and nutrients.
2. **Stimulated Microbial Life:** Trace minerals in seawater boost microbial activity, vital for breaking down organic matter and enriching soil fertility.
3. **Resilient Crops:** Plants treated with seawater minerals exhibit greater resistance to pests, diseases, and environmental stress.
4. **Sustainable Fertilization:** By reintroducing natural minerals, seawater reduces the reliance on synthetic fertilizers, aligning with eco-friendly practices.

Balancing Challenges

While seawater offers immense potential, its application must be approached carefully. Excessive salinity can harm soil and plants, emphasizing the importance of proper dilution and monitoring. Similarly, the by-product from desalination requires careful management to avoid unintended environmental impacts.

Seawater's profound connection to life extends beyond the ocean. It is enriched by the minerals of the land, sustains our bodies, revitalizes our soils, and holds the key to sustainable agriculture. By embracing innovative practices that repurpose seawater and desalination by-products, we can harmonize with nature's design, ensuring that the ocean's bounty nourishes both humanity and the earth for generations to come.

HARNESSING WATER'S FIRE

THE TRIPLE POINT OF WATER IN THUNDERSTORMS & VOLCANIC ERUPTIONS

In nature, high-energy phenomena like thunderstorms and volcanic eruptions reveal an awe-inspiring process: water reaching its triple point. At this state, water exists simultaneously as a solid, liquid, and gas, unleashing vast amounts of energy. Far from a mere metaphor, this convergence of phases embodies the dynamic power within water's fundamental structure. It serves as a reminder that water, under specific conditions, becomes an engine of nature itself, creating life-sustaining forces through dramatic releases of energy.

The Triple Point in Action: Thunderstorms and Volcanic Eruptions
In thunderstorms, as water ascends and temperatures drop, it forms ice, liquid droplets, and vapor in the same region of the atmosphere. These conditions create powerful electrical charges, leading to lightning–the visible force generated by water's triple-point energy. Similarly, volcanic eruptions force water trapped underground into extreme temperatures and pressures, causing water to reach its triple point and vaporize explosively. This transition drives eruptions, propelling magma, ash, and gases into the air with a force that shapes landscapes and ecosystems.

Harnessing the Triple Point for Energy
By observing these natural events, we glimpse the extraordinary potential of water as it reaches its triple point, where ice, water, and vapor coexist in a delicate yet powerful balance. This phenomenon serves as a direct example of how water's properties can create sustainable, high-energy systems. In designing engines or energy sources inspired by the triple point, we could emulate the forces in thunderstorms or volcanic eruptions, accessing energy without the need for destructive combustion.

Imagine a vehicle engine capable of inducing water's triple-point conditions on a controlled scale. This engine would draw from nature's own processes to create energy in a balanced, i**mplosive way rather than the explosive methods of traditional combustion. Using water at its triple point, this approach taps into the natural forces seen in thunderstorms and volcanic eruptions, unlocking clean, potent energy.**

The Three Elements in Motion: Water, Air, and Fire
In the context of this book, water, air, and fire converge at the triple point. Water provides the core medium, air facilitates the phase transitions, and the intense conditions create a fire-like release of energy. In this state, water becomes a force of transformation–a natural engine that illustrates the potential for sustainable power inspired by nature's own mechanisms.
Honoring Nature's Forces for Sustainable Innovation

Thunderstorms and volcanic eruptions demonstrate that the energy potential within water is not simply theoretical; it's a natural phenomenon that has shaped ecosystems and landscapes for eons. By harnessing these processes at the triple point, we can explore new paths to clean energy–working in harmony with nature rather than extracting from it. This perspective invites us to see energy not as something to burn or exploit, but as a balanced force to coax and channel, guided by nature's rhythms and the forces within water itself.

This image depicts the full cycle of water at its triple point, showing the interconnected stages from the underground aquifer to volcanic eruption, thunderstorm, and rainfall. It captures the dynamic flow of water through its various phases—solid, liquid, and gas—in a single, powerful scene.

This triple point represents a powerful cycle of energy and life. It shows how different forces come together, creating something more significant than their individual contributions, similar to an idea evolving from imagination to a tangible form. In its fluid phase, an idea flows freely, taking shape through creativity. With passion–our inner "fire"–the idea gains momentum, shifting into an active state. Finally, in its solid phase, the idea materializes and becomes a reality.

THE MYSTERY OF PRIMARY WATER

EARTH'S HIDDEN LIFEBLOOD

Primary water is an often-overlooked but profoundly significant aspect of the Earth's natural cycles. Unlike surface or rainwater that cycles through rivers, lakes, and oceans, primary water originates deep within the Earth, formed through intense geological processes involving heat and pressure. Hydrogen and oxygen combine within the Earth's mantle to create this pristine water, which rises to the surface through fractures and faults in the crust, independent of the hydrological cycle driven by precipitation.

The Bible references the "waters beneath the Earth," aligning with ancient traditions that recognize primary water as a sacred and life-giving source. These waters were often revered in spiritual texts, highlighting their connection to creation and renewal. Modern mining operations and geological studies have also revealed springs of primary water, particularly in arid regions where surface water is scarce, demonstrating its untapped potential and resilience.

Accessing primary water requires both ancient wisdom and modern technology. Natural springs where primary water surfaces on its own have been used by indigenous cultures for centuries, often regarded as sacred sites. Today, deep-drilling wells, guided by advanced geological mapping and seismic imaging, allow mankind to tap into these hidden reservoirs. These wells target deep aquifers formed by primary water, not replenished by rainfall but created through Earth's internal processes. Such methods demonstrate that accessing primary water is not a matter of scarcity but one of knowledge and awareness.

For centuries, dowsing rods have been used to locate underground water sources, including primary water. Dowsing involves holding L-shaped metal rods or a forked branch, such as one from a willow or hazel tree, and walking over the land to sense subtle Earth energies. The rods are believed to move or cross when water is detected underground, often at geological fractures or fissures where primary water flows.
Dowsing is deeply rooted in ancient traditions, practiced in Egypt, China, and Europe as a sacred skill that aligns with Earth's electromagnetic field. Dowsers are thought to have an enhanced sensitivity to the natural forces that guide the flow of water. This practice bridges the mystical and the practical, allowing humanity to tap into unseen forces to uncover Earth's hidden lifeblood.

Primary water is pure and untainted by surface contaminants, holding the essence of Earth's primordial vitality. Its discovery and use can transform regions suffering from water scarcity, offering a sustainable solution deeply rooted in the planet's natural processes. This water reminds us of the profound interconnectedness of Earth's layers and humanity's sacred relationship with the hidden forces of nature.

Nature itself seeks balance. Just as the human body is composed of approximately 70% water, so too is the surface of the Earth. This parallel invites us to ponder what lies below–the hidden reservoirs and cycles that mirror our internal composition. Beneath the Earth's crust, primary water flows in a quiet rhythm, embodying the same principles of regeneration and sustenance found within our own bodies. It is Earth's "lifeblood," coursing through its veins and sustaining its equilibrium, a profound reflection of the interconnectedness of all living systems.

The question then arises: do we truly face a water shortage, or are we suffering from a lack of past wisdom and understanding of Earth's regenerative systems? Primary water challenges the modern perspective of scarcity by revealing the planet's ability to create and sustain life in ways often overlooked. Exploring primary water invites us to reconsider our relationship with natural resources, to seek harmony with the Earth's rhythms, and to respect the profound wisdom encoded within the planet itself. By embracing this knowledge, we can ensure the preservation and equitable use of this invaluable resource for generations to come.

WATER AS A MIRROR TO HUMAN CIVILIZATION

Schauberger believed water's health mirrored the health of human society, suggesting that polluted or "dead" water was a reflection of society's disconnect from nature. This aligns with the idea that environmental degradation is a sign of spiritual and cultural imbalance, a philosophy that informs modern environmental and sustainability ethics.

His philosophy encourages humanity to view water as a teacher and guide, suggesting that by observing and honoring water's role in ecosystems, we can restore balance within ourselves and our communities.

DANGERS OF OUR INDUSTRIAL AGE

PRESERVING EARTH'S LIFEBLOOD

Before the Industrial Age, all water on Earth maintained a structured, harmonious state, aligning with nature's rhythms and nurturing life. As humanity has intervened in natural processes, tampering with the delicate ecosystems of water and soil, the balance has been disrupted, leaving these elements less able to support life in their original vitality. Structured water, when left to its natural condition, even carries bio-photonic energy—a form of self-sustaining light that is fundamental to health and growth.

However, when water is exposed to toxic chemicals, residues from industrial runoff, and synthetic materials, it remembers and stores these foreign vibrations deep within its molecular structure. This stored toxicity impacts not only the environment but also the ecosystems relying on pure, structured water. Household cleaners, often filled with harsh chemicals, are especially impactful; they dry the skin just as they dry the land, leaving a "desert" effect on the earth's surface.

Consider: Avoid storing chemicals under your sink, as water absorbs vibrational information from its surroundings. When exposed to chemicals, even in sealed containers, water can carry subtle energetic imprints from them. Through this process, we may unknowingly absorb these vibrations into our own system, affecting our body's natural balance. Keeping your water's environment pure helps maintain its vitality and aligns with our health and well-being.

"*Water is a living substance.*" ~ Victor Schauberger, Naturalist and Inventor

CHAPTER 7

SACRED CURRENTS

THE SACRED MIRROR

Water, in its tranquil embrace, becomes a sacred mirror, reflecting not only the world around but the depths of our very souls. As we gaze into its stillness, we meet not just our image but the whispers of our thoughts, the echoes of our emotions, and the stories etched within us.

In nature's gentle hand, water serves as a canvas that captures the ever-changing sky, the fiery blush of sunsets, and the lush greenery of life. It mirrors beauty and chaos alike, inviting us to pause, to breathe, and to contemplate our place within the grand tapestry of existence. Each ripple dances with the light of the sun, beckoning us to explore the depths beneath the surface.

"Just as water reflects the light from above, our lives reflect the energy we give out into the world. The feedback we receive shapes our journey." ~ Unknown

Water is not just a passive mirror; it is a holographic recorder, a living matrix that holds the light of all that we are. Within its crystalline structure, the light of our being is imprinted, recorded like a memory etched into the fabric of the universe. Water reflects the light of our thoughts, our feelings, and our intentions, storing this information in its molecular structure. Just as a hologram captures the essence of an object in three dimensions, water captures the essence of our consciousness, holding the imprint of who we are, reflecting our deepest truths.

"Water, like a hologram, contains the essence of all that it reflects. It holds not just the image but the energy, the light, and the life force itself." ~ Unknown

This recording is akin to the Akasha records, a concept of an invisible, ethereal plane where the collective memory of the universe is stored. The Akasha, often referred to as the "cosmic library," holds the light and the records of all beings, events, and experiences. Just like a hologram or the Akasha, water serves as a vessel for light, recording the energy, vibrations, and information that flow through it.

"As the Akasha records the stories of the soul, water records the pulse of the universe, reflecting the imprint of every thought, feeling, and action." ~ Unknown

Water, in its boundless capacity, acts as both reflector and recorder, offering us feedback about our state of being. Just as light moves through its matrix, so too do our energies and emotions create patterns that the water records, providing us with an ongoing feedback loop. In this dance of reflection and record, water becomes a sacred ally, showing us the imprints of our own light as we move through the world. It holds our essence and offers us guidance, acting as a bridge between the conscious and the unconscious, providing us with an ever-present mirror of our inner world.

"Water, the universal solvent, carries the imprints of all that it touches, storing the light of the cosmos like an eternal memory."

To look into water is to seek clarity; the calm surface reveals reflections as clear as our deepest truths. In this sacred element, we find purification and renewal, a reminder that like water, we too can cleanse our minds and spirits, embracing the tides of transformation. This is the feedback loop: we release, we reflect, and we evolve.

In a world that often rushes by, the stillness of water calls us to slow down, to reflect. Each glance into its depths invites us to dive deeper, exploring the currents that flow through our lives. Every ripple signifies a thought, an emotion—each a dynamic interplay between our inner and outer worlds.

As we recognize water as our mirror and recorder, we glimpse our connection to the cosmos. Just as the surface of a pond reflects the stars above, we, too, are woven into the fabric of the universe. Our emotions and experiences ripple outward, creating waves that touch the shores of existence and all those around.

"The water mirrors the stars, as the soul mirrors the universe. In each drop, the infinite is reflected, and in each ripple, the pulse of life is felt." ~ Unknown

In embracing the essence of water as a mirror and recorder, we open ourselves to its wisdom. We learn to reflect upon our journeys, honor our true selves, and navigate the currents of life with grace and awareness. Through its feedback loop, water not only shows us who we are but helps us evolve into who we are becoming, as it records our light in its endless, flowing matrix.

Water as the universal solvent, carrying the imprints of all that it touches and storing the light of the cosmos.

Reflected in the water, glowing with light to symbolize consciousness and universal energy.

THE JOURNEY OF WATER

STAGES OF TRANSFORMATION

As we move through life, we experience moments of crystallization, where certain experiences, ideas, and emotions freeze in clarity, leaving imprints on our inner world. These are moments that define and shape us, like water molecules aligning in formation, creating inner structures that reflect who we are becoming.

As water moves and changes, so do we. Life brings moments of intense warmth, like sunlight on the sea, where we feel our form evaporate, releasing parts of ourselves back to the source. These are times of transformation, when we shed past identities, letting go of patterns, memories, and even parts of our physical selves. Evaporation allows us to return to a formless state, where we are pure potential once again, merging back into the vastness, preparing for another cycle.

This process mirrors the journey of water: the drop evaporates, rises, gathers, and condenses, only to return as rain, rejoining the ocean. So, too, do we undergo cycles of creation, release, and renewal, returning to our essence, to the whole, while forever carrying within us the memory of each experience. We are individual yet connected, crystallized yet fluid, shaped yet ever-transforming.

In honoring this cycle, we see that our lives, like water, are part of a greater flow–a dance of emergence, transformation, and return. In each drop, in every phase, we are never separate but always part of a vast, boundless ocean that connects and sustains all.

"The only way life could leave the ocean was to take the ocean with it." ~ Rachel Carson

Water, in all its moods, romances the world with its essence, reflecting back the passions, stillness, and mysteries of the heart. Just as we live, love, and transform, so does water–forever changing, forever flowing, whispering to us the poetry of its own alchemy.

WATER'S SOUL

Tormented Waters
In the wild surge of tormented waters, there is a haunting beauty–an ocean's fury, like a tempest of emotions let loose. As waves rise and crash, they echo the depths of sorrow, anger, and passion, revealing a soul that cannot contain its intensity. Tormented water is poetry in motion, a symphony of chaos, where each crest and trough speaks of the heart's unguarded, primal cry.

Placid Waters
Picture a lake, its surface smooth as glass, untouched by wind or worry. Placid waters hold a timeless grace, a gentle stillness that invites peace into the heart. It is the quietude of love deeply felt and serenely accepted, a reminder that even in silence, there is connection, a quiet but unwavering bond that cannot be broken. In these waters lies the beauty of simplicity, a love so pure that it rests undisturbed.

Calm Waters
Calm waters flow as a lover's hand might trace gentle lines, moving with a soft, rhythmic grace. This water is a caress, an invitation to be soothed and embraced. Calmness glides like a river on a summer's day, filled with soft light and quiet joy. It brings a feeling of belonging, a heart at peace, moving in harmony with the world, content to drift, unfettered and free.

Reflective Waters
Gazing into reflective waters is like looking into the soul's mirror, a glimpse of truths often hidden. This water reveals, as a lover might, with honesty and care, laying bare all that we are, our beauty and our flaws alike. In its depths, we find clarity; we see ourselves anew, as reflections woven with light and shadow, vulnerable and beautiful in our entirety.

Charged Waters
Charged waters pulse with energy, sparkling under the sun or rushing like a wild river, alive with the thrill of existence. Here, water embodies passion, the intensity of a heart on fire. It is electrified by life itself, infused with purpose, an unstoppable force. Charged water, like love in full bloom, refuses to be held back, surging forward with an unbreakable vitality, illuminating everything it touches.

WHISPERS IN THE WATER

YOUR WORDS ARE YOUR SPELL

The words you bathe in are more than sound; they are the "solution" you absorb and swim in, an invisible elixir shaping your mind and body. The true solution to life's questions lies within the field of vibration we allow in, within the conversations we embrace, and the energy we carry. Our words become directives, sending commands that echo in every cell, as though casting a spell for our reality to follow.

When you speak, you formulate your life's work. Your words are the blueprint and genesis of your creation, the electric charge that resonates in the waters within you, magnifying your intent. You are both electric and magnetic–your words align, direct, and empower. Water, as the recorder of life, carries this resonance, imprinting every command, every thought, and every memory. This is why laughter can heal; joy vibrates at a frequency that water gladly receives, magnifying life.

"I have the solution of the universe rushing through me.
It holds my memories, like a tape recorder embedded in my being,
playing back my words and thoughts, reminding me of my power.
I choose the waters I swim in, the words I let sink in,
and the words I send forth into the world.
The loving word is the only true solution,
bringing peace and resolution within."
- Karen Elkins

Judas said, Tell us what existed before heaven and earth came into being. "The Lord said, *"There was darkness, and water, and the Spirit came upon the water. Yet I tell you truly that what you are seeking is already within you..."* Dialogue with the Savior

They both listened silently to the water, which to them was not just water, but the voice of life, the voice of Being, the voice of perpetual Becoming." — Hermann Hesse, Siddhartha

THE BREATH OF WATER

DIVINE ESSENCE AND THE EXCHANGE

In this delicate balance, we see ourselves as both carriers and caretakers, each breath part of an ancient, sacred flow. The term "sacred" refers to something regarded with reverence, embodying profound respect and connection to the divine or the essence of life itself. Water purifies and sustains, embodying life's cycles, while air cradles us, shaping our energy and well-being. We are deeply intertwined with this continuous flow of air, water, and earth–a web of life in perpetual motion, forever connected to the rhythm of the universe within and around us.

Every breath we take is an act of sacred connection–a moment where we become one with the world. It is more than just oxygen; it is an intricate blend of water vapor, particles, and life drawn from the farthest reaches of the earth. This breath carries the essence of forests, rivers, mountains, and seas, reminding us of our intimate bond with nature's pulse. There is a rhythmic and sacred exchange from us to the trees, a cycle of giving and receiving that sustains life and underscores the unity between humanity and the natural world.

Trees gift us the exchange of life-giving oxygen and take in carbon dioxide in return, creating a vital balance that sustains life. This cycle exemplifies the harmony of elements we share–oxygen, carbon dioxide, and water vapor–essential components of the breath that connects all living beings. Additionally, the ocean, which covers the majority of our planet, provides approximately 75% of the oxygen we breathe. Through the intricate dance of phytoplankton and marine ecosystems, the ocean becomes an unseen yet powerful force, contributing to the continuous flow of air that nourishes us and links us to the entire web of life.

"Man is not just a product of the earth; he is also the earth's caretaker. We are connected to the cosmos, and the divine breath flows through us in harmony with nature." - Rudolf Steiner

"Water is the carrier of life and the pulse of the planet. It purifies, nourishes, and sustains the sacred balance that supports all existence.

"Comprehend and copy Nature; that is where our salvation lies."
- Viktor Schauberger

"Science is not only compatible with spirituality; it is a profound source of spirituality. When we recognize our place in an immensity of light-years and in the passage of ages, when we grasp the intricacy, beauty, and subtlety of life, then that soaring feeling, that sense of elation and humility combined, is surely spiritual. So are our emotions in the presence of great art or music or literature, or acts of exemplary selfless courage such as those of Mohandas Gandhi or Martin Luther King, Jr. The notion that science and spirituality are somehow mutually exclusive does a disservice to both." - Carl Sagan, *The Demon-Haunted World: Science as a Candle in the Dark*

"All motion is a universal heartbeat, pulsing forever in the cosmic sea of living energy." - Walter Russell

WHO AM I?

Your journey is one of co-creation with the Force of all existence. This Source is within you and all around you. As a seeker, you can travel seeing purely the physical or cross over and explore the mystical. The hero's journey is to "re-discover" the magnificence of who you are and to "re-cognize" the divinity within.

"What am I? And the Voice answers: "I am an expression of the universal passion of creation. "God created me that I should fulfill His purpose. "God gave me the desire to create and the power of creation. "God dwells within me. I shall not deny the power within me, which is God within me.

"I shall not close the ears of my soul to the whisperings of my soul, which make me dwell on the mountain top in the ecstasy of inner thinking.

"The universal desire is the expression of an idea through the rhythm of thinking, in accord with the law, in endless sequence throughout endless space." ~ Walter Russell, The Universal One

Source has given us a toolbox from which to create and understand our elemental selves. Explore and become mindful of the chemistry that cascades within. Become the master of these elements—thought, emotion—and do not let your senses be the sole ruler of your destiny. Tap into the cosmic mind—ask, and the door shall be opened to cosmic wisdom.

CHAPTER 8

SACRED WATERS

SACRED WATERS'

WATER AS LIFE'S ESSENTIAL MEDIUM AND HEALER

"In every drop of sacred water lies a reflection of the cosmos, a conduit for divine essence that flows through all of life."

Sacred water holds a revered place across cultures, symbolizing purity, transformation, and connection to the divine. Known as "holy water" in many traditions, it's seen as a medium that cleanses not just the body but also the soul, acting as a bridge between the earthly and the spiritual. Whether from rivers, springs, or ritual vessels, sacred water is imbued with blessings, prayers, and intentions, elevating it to a medium for spiritual renewal.

Water's ability to hold memory and vibrational energy adds a deeper layer to its sacred role.
From the Ganges in India to Lourdes in France, holy waters are believed to contain unique energies capable of healing, purifying, and awakening one's higher self. In rituals, these waters often symbolize rebirth, connecting people to a collective spiritual consciousness that transcends physical boundaries.

SACRED SITES

SPACES OF REVERENCE AND TRANSFORMATION

Water, one of the most essential elements of life, has long been recognized not only for its physical properties but also for its spiritual and energetic significance. Across ancient civilizations, from the temples of Egypt to the stone circles of Britain, water played a central role in creating spaces of reverence and transformation. Many of these sacred sites were intentionally aligned along what Freddy Silva, in his book Divine Blueprint, refers to as byelines and natural water channels, harnessing the earth's energy flows to amplify the sites' spiritual resonance.

"Water flows beneath many sacred sites, serving not only as a physical source of life but as a channel for earth energies, enabling deeper communion with the spirit." — Freddy Silva

Byelines, also known as ley lines, are thought to be lines of energy running through the earth, connecting places of historical and spiritual importance. When a site is built along these lines and near flowing water, it connects with a powerful, dynamic energy circuit. Water channels, often beneath the surface, bring energy from the depths of the earth, symbolizing a direct link to life's origins. These channels not only physically hydrate the land but are believed to conduct subtle earth energies that heighten the effects of the byelines.

"Throughout history, water and sacred geometry have intertwined to create spaces where life and spirit converge."
— Freddy Silva

At sacred sites, water channels serve as conduits for this energy, providing a kind of sacred hydration that nourishes the earth, the structures above, and those who visit. Ancient builders understood that by aligning their temples, stone circles, and monuments with these energy pathways, they could create spaces where spiritual insight, healing, and communion with the divine were more accessible.

Water and Sacred Geometry: The Blueprint of Life

Water's presence at these sites wasn't coincidental–it followed a blueprint that aligned with nature's own geometry. Just as water's molecular structure mirrors geometric principles found in sacred architecture, the placement of these sites often aligned with patterns that harnessed natural resonance. By placing sacred sites along channels and byelines, ancient cultures enhanced the sites' ability to facilitate healing and spiritual awakening. Water became not just a symbol of life, but a vehicle of divine connection.

The Legacy of Sacred Water

Today, the alignment of sacred sites with water channels and byelines continues to inspire those who study these places and seek to understand the natural forces behind them. They serve as a reminder that water is more than a physical substance–it is the life force that connects and amplifies all things. When we step onto these sites, we are immersing ourselves in a harmony of earth, water, and spirit, tapping into an ancient understanding of how to align with the natural and sacred rhythms of life.

THE MIRROR OF LIFE

SPIRITS OF LIGHT, AND THE CANOPY OF CREATION

Water is the mirror of life. It reflects our inner world, connects us to other spirits of light, and bridges us to the vastness of creation. Within its depths and surfaces, water holds the stories of existence, reflecting back to us who we are and who we might become. It is more than a substance; it is our canvas, an endless pool where the colors of life swirl, blending the physical with the spiritual.

At its essence, water holds the full spectrum of light, a palette from which our emotions and experiences are painted. Each droplet captures shades of joy, sorrow, love, and awe, while carrying the full range of cascading chemistry that runs through our veins. This elemental flow links us to the earth and to one another, grounding us within the natural world while carrying the spark of life within.

Like the intricate mosaics in cathedrals and mosques, where vibrant tiles form the majestic domes that open toward the heavens, water is here to remind us of life's kaleidoscope. In its endless reflections, water invites us to see the beauty of diversity, to witness each spirit as a unique expression of the universal light. Just as every tile in a grand dome contributes to a breathtaking whole, each of us is part of a larger, vibrant mosaic, a tapestry of beings connected by the flow of water, light, and life.

In the great dance of existence, water is our partner, reflecting back our passions, dreams, and innermost truths. It offers a canvas to explore the full spectrum of life, inviting us to immerse ourselves in its depths, to trust in the flow that connects all, and to experience the living chemistry that mirrors the very elements running through our own veins. Through water, we are guided to witness the majestic mosaic of life, to create, to reflect, and to remember that we are bound by light, forever part of life's eternal mirror.

"Within the sanctuary of light and form, I am both observer and observed, immersed in the beauty of existence reflected back to me."

"In the vastness of creation, I am but a thread—yet within this sacred space, I am whole, connected, and woven into the tapestry of light."

SACRED WATER RITUALS

PURITY, RENEWAL, AND SPIRITUAL CONNECTION

Sacred water rituals, such as baptism, purification, or anointing, hold profound significance across various cultures and spiritual traditions. Water, often regarded as the essence of life, is not only a physical substance but a symbolic representation of purity, renewal, and spiritual connection. Through these rituals, water transcends its material form, becoming a conduit for divine presence and transformation.

In the Christian tradition, the washing of feet by Jesus symbolizes humility, servitude, and the profound act of cleansing not just the body, but also the soul. In this ritual, water becomes a sacred vessel, carrying the energy of compassion and service. It is a moment of connection, where the act of physical cleansing reflects a deeper purification of the spirit, encouraging unity and empathy among individuals.

Across the world, in cultures ranging from ancient India to indigenous traditions, water has been used as a sacred tool for spiritual transformation. The flowing of water symbolizes the constant flow of divine energy, washing away impurities, both physical and spiritual. It is through this sacred engagement with water that individuals are reminded of their connection to the divine and the unity of all living beings. Water, in this sense, acts as a bridge between the material and spiritual realms, facilitating not only physical renewal but also emotional and spiritual growth.

The act of immersion or anointing with water is a ritual of rebirth–a cleansing that resets the soul, renewing it for a higher state of being. Through the symbolic use of water, people are reminded that they are part of a greater cosmic flow, interconnected with all forms of life. Water's sacredness lies not only in its practical role in sustaining life but also in its ability to symbolize the divine purity and transformation that flows within each of us.

In embracing these water rituals, we recognize the universal power of water as a divine substance that can heal, purify, and unite us all. By engaging with water's sacred essence, we tap into an ancient tradition that connects us to the deepest layers of existence, to the very pulse of creation.

Sacred water rituals, whether involving baptism, purification, or anointing, remind us of the universal flow that connects and revitalizes all beings. By engaging with water's sacred essence, we embrace a tradition that venerates water as a divine substance, holding space for spiritual transformation and unity across cultures. Would you like an image that visually captures this sense of water as sacred and holy?

HOT SPRINGS

STORE, TRANSMIT, AND AMPLIFY HEALING ENERGIES

Hot springs are like nature's alchemical baths, harnessing the Earth's energy to provide a profoundly nourishing experience. The minerals in these waters don't merely dissolve–they interact with water's unique ability to store, transmit, and amplify healing energies, creating a therapeutic synergy unmatched by any other element.Hot springs, often referred to as nature's alchemical baths, offer a profound healing experience by drawing on the Earth's energy and mineral richness. These naturally heated waters create a unique synergy of therapeutic benefits through their physical, chemical, and energetic properties.

1. Mineral-Rich Composition

Magnesium: Relaxes muscles, calms the nervous system, and supports heart health.
Calcium: Strengthens bones and supports cellular communication.
Sulfur: Known as the "beauty mineral," it helps detoxify the body and improve skin health.
Lithium: Found in some hot springs, it can help stabilize mood and reduce anxiety.
Sodium and Potassium: Maintain electrolyte balance, enhance hydration, and support nerve function.

2. Thermal Energy and Circulation

The heat from hot springs improves blood circulation by dilating blood vessels, which enhances oxygen and nutrient delivery throughout the body.

Improved circulation helps to detoxify tissues and organs, supporting overall vitality.

3. Stress Relief and Relaxation

The soothing warmth reduces muscle tension and promotes relaxation, releasing endorphins that elevate mood and ease chronic stress.

4. Skin Health

Minerals like silica and sulfur rejuvenate the skin, promoting elasticity, reducing inflammation, and helping with conditions like eczema, acne, and psoriasis.

Regular soaking can improve the skin's texture and clarity, acting as a natural exfoliant.

5. Detoxification

The combination of minerals and heat helps open pores and flush toxins from the body, aiding lymphatic drainage and reducing water retention.

6. Pain Relief and Joint Support

Hydrotherapy from hot springs has been shown to alleviate pain associated with arthritis, fibromyalgia, and other musculoskeletal conditions.

The buoyancy of water reduces pressure on joints, making it ideal for gentle movement and therapy.

7. Energetic and Subtle Healing

Hot springs often carry Earth's energetic charge, amplified by water's unique ability to store and transmit energy. This can have a harmonizing effect on the body's biofield, restoring balance and vitality.

Some hot springs, especially those with high geothermal activity, are believed to be centers of electromagnetic energy, enhancing healing at a vibrational level.

8. Immune Support

Regular immersion in hot springs can stimulate the production of white blood cells, enhancing immune function.

Sulfur's detoxifying properties further bolster the immune system by reducing the body's toxic load.

9. Connection to Nature

The holistic benefits extend beyond the physical, as hot springs provide an opportunity to reconnect with nature. This connection fosters mental clarity, emotional balance, and spiritual grounding.

10. Hydrotherapy's Historical Legacy

Many ancient cultures, from the Romans to the Japanese, have recognized and utilized hot springs for their therapeutic properties, embedding them in traditions of healing and community well-being.

ONE SOURCE, ONE SCIENCE

JEREMIAH 2:13 AND 17:13, PORTRAYING GOD AS "THE SPRING OF LIVING WATER."

The concept of God, or a divine source, as "the spring of living water" appears in many spiritual traditions worldwide. Here are some examples:

Christianity

John 4:14: Jesus refers to Himself as the source of "living water," offering eternal life to those who believe in Him. Revelation 22:1: The vision of the *"river of the water of life,"* flowing from the throne of God and the Lamb, represents divine grace and eternal life.

Judaism

Psalm 36:9: *"For with You is the fountain of life; in Your light, we see light."* This verse parallels the idea of God as the source of life and spiritual renewal.

Islam

Qur'an 21:30: *"We made from water every living thing."* Water is deeply symbolic of life and creation in Islam, representing divine mercy and the sustaining power of Allah.

Hinduism

Rigveda: The Vedic texts describe rivers and water as sacred and life-giving, often seen as manifestations of divine power (e.g., Saraswati, the goddess of a sacred river).

Native American Traditions

Water is life. Without it, we cannot live. We are its keepers, and it is our sacred duty to honor and protect it."– Hopi teaching

Taoism

Tao Te Ching (Chapter 8): Laozi likens the Tao to water, saying, "The highest good is like water," as it sustains all things without striving. Water symbolizes humility, persistence, and adaptability, qualities of the divine order.

Zoroastrianism

Zoroastrian Prayer to Water: "We worship the waters, the good, the holy, the Mazda-created waters."

Websites to Products to Heal our Planet

Chapter 2 - Waters Secrets

- **Albert Szent-Györgyi- Books and Site**
  https://www.sciencehistory.org/education/scientific-biographies/albert-szent-gyorgyi/

1. *Bioenergetics*
2. *Introduction to a Submolecular Biology*
3. *Electronic Biology and Cancer: A New Theory of Cancer*
4. *Bioelectronics: A Study in Cellular Regulations, Defense, and Cancer*

- **Rene Quinton** - waterandwellness.com (Quinton Products)

1. *L'eau de mer, milieu organique*
2. *René Quinton y el agua de mar: Toda su vida, su historia y sus aplicaciones con el agua de mar*

- **Dr. Gerald Pollack - Books and Site**
  pollacklab.org

1. *The Fourth Phase of Water: Beyond Solid, Liquid, and Vapor*
2. *Cells, Gels, and the Engines of Life*
3. *Muscles and Molecules: Uncovering the Principles of Biological Motion*

- **The Water Bridge Experiment**
  https://www.researchgate.net/publication/231015252_The_floating_water_bridge

Chapter 3 - Water's Hidden Realm: The Science Within

- **Dr. Mae-Wan Ho -Books**

1. *Living Rainbow H2O,*
2. *Meaning of Life & the Universe*
3. *The Rainbow and the Worm, the Physics of Organisms*

- **Dr. Esther Patricia Perez Del Rio** - reinnoa.com
- **Nikola Tesla**
- **Dr. Luc Montagnier - Nobel Prize-winning French virologist, books**

1. *Virus: The Co-Discoverer of HIV Tracks Its Rampage and Charts the Future*
2. *Les Combats de la Vie (The Battles of Life)*
3. *L'Ea u et la Physique Quantique (Water and Quantum Physics)*
4. *Les Chemins de la Vaccination (The Pathways of Vaccination)*

   http s://www.nobelprize.org/prizes/medicine/2008/montagnier/facts/

- **Prof. Dr. Bernd Kröplin- Books and Site**
  https://www.weltimtropfen.de/index_english.html

1. *Water: Much More Than H2O*
2. *World in a Drop: Secrets of Water*
3. *Water Drops: Images of the Hidden World*

- **Dr. Masaru Emoto** - Books and Offical Site masaru-emoto.net

1. The Hidden Messages in Water
2. The True Power of Water: Healing and Discovering Ourselves
3. Messages from Water and the Universe
4. The Secret Life of Water

- **Dr. Konstantin Korotkov** - Books & Site

1. www.b-wellnow.net/dr-korotkov.html
2. Light After Life: Experiments and Ideas on After-Death Changes of Kirlian Pictures
3. Human Energy Field: Study with GDV Bioelectrography
4. Energy Fields Electrophotonic Analysis in Humans and Nature
5. Science of Measuring Energy Fields
6. The Energy of Space: Scientific and Practical Aspects of Bioelectrography
7. The Energy of Health: Diagnosing and Optimizing Human Energy

Chapter 4 - Crystalline Lattice, Table of Elements and the Color of Life

- **Randy Hatton**

1. vibrantvitalwater.com

- **Walter Russell** - Books & Site

1. philosophy.org
2. Universal One
3. The Secret of Light
4. The Message of the Divine Iliad
5. A New Concept of the Universe
6. Atomic Suicide?

- **Don Estes**
- Innersense-inc.com

- **Gary Robert Buchanan** - Books and YouTube

1. Book: Sona: Healing with Wave Front Bioresonance
2. youtube.com/@sonatherapy/videos

- **Samuel Hahnemann**

1. https://www.cureus.com/articles/289508-the-life-and-legacy-of-samuel-hahnemann-founder-of-homoeopathy-and-his-medical-philosophy#!

Chapter 5 - Vibration, Light, Evolution

- **John Stuart Reid** - cymascope.com

- **Richard Merrick** - Interferencetheory.com

1. *Interference: A Grand Scientific Musical Theory*
2. *The Venus Blueprint: Uncovering the Ancient Science of Sacred Spaces*

- **Walter Russell** - philosophy.org

1. *Universal One*
2. *The Secret of Light*
3. *The Message of the Divine Iliad*
4. *A New Concept of the Universe*
5. *Atomic Suicide?*

- **Electrical Discharge** (Electric Universe Model) - thunderbolts.info

- **Bio-Geometry** - biogeometry.ca

1. *Back to a Future for Mankind: BioGeometry*
2. *BioGeometry Signatures: Harmonizing the Body's Subtle Energy Exchange with the Environment*
3. *Hidden Reality: The BioGeometry Physics of Quality*

Chapter 6 - Water the Work-horse of Nature

- **Viktor Schauberger** - (6/30/25-9/25/58) - https://pks.or.at/en/

1. The *Water Wizard: The Extraordinary Properties of Natural Water*
2. *Nature as Teacher: New Principles in the Working of Nature*
3. *The Fertile Earth: Nature's Energies in Agriculture, Soil Fertilisation and Forestry*

- Dr. Krishna Madappa - krishnamadappa.com. and jivawater.com

- **BioMimicry** - biomimicry.org

1. Bio*mimicry: Innovation Inspired by Nature*
2. *Biomimicry in Architecture*
3. *The Shark's Paintbrush: Biomimicry and How Nature is Inspiring Innovation*

- **George Wiseman** -Eagle-research.com

1. Hydrogen Machines

"We are but whirlpools in the cosmic river of life. Each twist, turn, and ripple is a reflection of the universal flow, as above, so below, so within."

- Dr. West Marrin Ph.D. - watersciences.org

1. *Altered Perceptions*
2. *Hydromimicry*
3. *Universal Water*

- The Branching of Life - Mandelbrot Set

1. *https://mathworld.wolfram.com/MandelbrotSet.html*
2. *The Fractal Geometry of Nature. - Benoit B. Mandelbrot*
3. *Infinite Patterns: The Mandelbrot Set and Chaos Theory - Clifford A. Pickover*
4. *Video - The Colours of Infinity: The Beauty and Power of Fractals*

Chapter 8 - Sacred Waters

- *Sacred Waters' - en.wikipedia.org/wiki/Sacred_waters*
- *Sacred Sites - Freddy Silva, Book Divine BluePrint*
- *Hot Springs - en.wikipedia.org/wiki/List_of_hot_springs*

The wave is the cosmic stage, the light is the actor, and the play is the drama of creation."

- Walter Russell

WHAT WE DO TO OUR PLANET, WE DO TO OURSELVES

"Ours is truly a water planet and that recognizing and respecting water's critical roles in processes such as climate change, energy selection, food production, economic stability, technology development, human health, and global conflict is not only wise, but indeed mandatory. Many of our recent choices have ignored the wisdom of both water and nature in adopting technologies and management schemes that violate, ignore, or just override water's natural rhythms, patterns, and preferences." ~ Dr. West Marrin

RESOURCES

The inspiration for this exploration stems from my magazine, Science to Sage, and my book, *InsideOUT: Revealing the Mysteries of Creation and the Wisdom to Live More Consciously* Connected. These works serve as tributaries merging into a greater journey–one that delves into the profound mysteries of water and its deep connection to the essence of life.

The images within this book, created through the innovative lens of AI, offer visual interpretations of water's story–an invitation to experience its beauty and transformative power from a fresh perspective. For those who wish to see these concepts reflected in real-world expressions, I warmly invite you to explore the water-themed editions of Science to Sage and my book, where photography and science come together to reveal the authentic patterns and wonders of water.

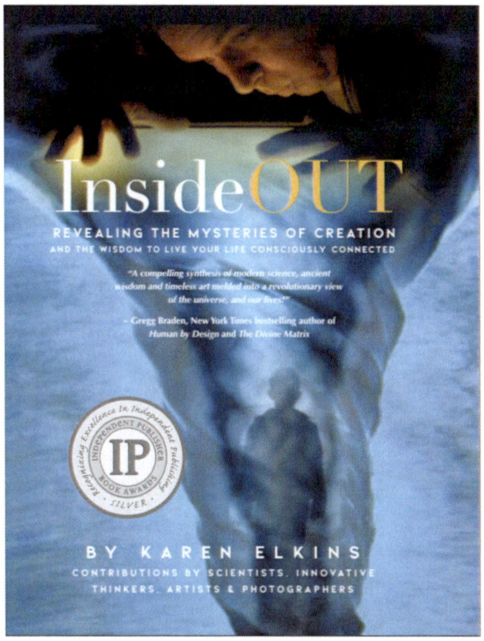

Sciencetosagemagazine.com
Find links to water issues and interviews

Connect:

Echosofwatter.com
SciencetoSage.com
Sciencetosagemagazine.com
Insideoutbook.com
www.amazon.com/Insideout-Revealing-Mysteries-Consciously-Connected

---

About Author:
Karen Elkins is the founder and editor of Science to Sage, e-magazine that bridges science, spirituality, and art. Known for her ability to weave together cutting-edge scientific insights, ancient wisdom, and creative expression, her work inspires a deeper understanding of the universe and the interconnectedness of life.

Karen is also the award-winning author of *InsideOUT: Revealing the Mysteries of Creation and the Wisdom to Live Your Consciously Connected.* This visually stunning and thought-provoking book has garnered global recognition for its exploration of sacred geometry, the elements of nature, and the profound patterns that underlie existence. Combining artistry, science, and philosophy, InsideOUT offers readers a transformative journey through the mysteries of creation.

Through Science to Sage and her writing, Karen collaborates with visionaries, scientists, and artists worldwide, making her a catalyst for innovative thought and a voice for the wonder of creation. Her work continues to inspire individuals to view the universe through a unified and awe-inspiring lens.

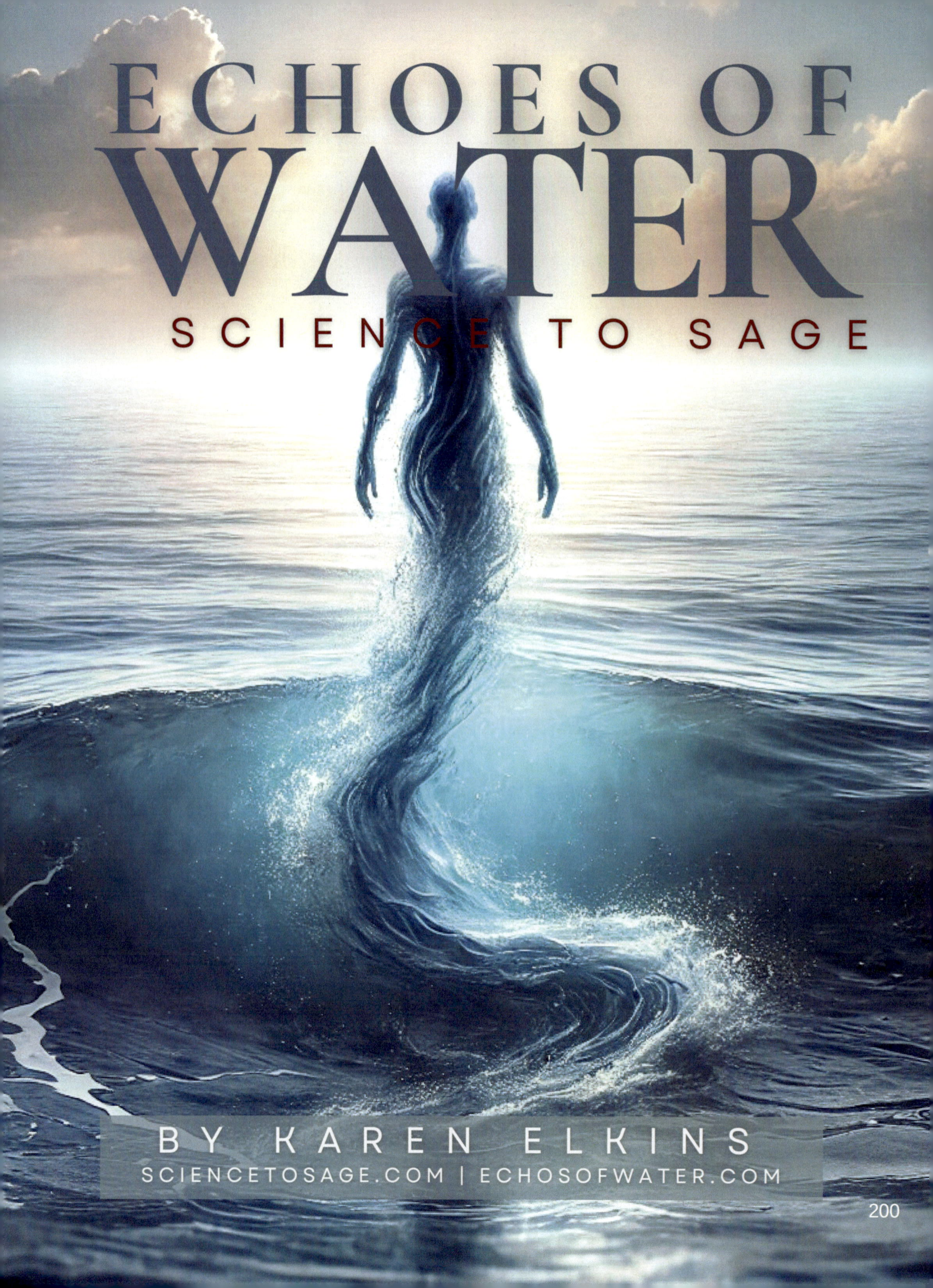
ECHOES OF
WATER
SCIENCE TO SAGE
BY KAREN ELKINS
SCIENCETOSAGE.COM | ECHOSOFWATER.COM

www.ingramcontent.com/pod-product-compliance
Lightning Source LLC
LaVergne TN
LVRC080247220926
848482LV00042B/344

* 9 7 9 8 3 4 8 2 7 8 3 9 7 *